# Sustainable Electricity

Jessica Fox
Editor

# Sustainable Electricity

Case Studies from Electric Power
Companies in North America

Springer

*Editor*
Jessica Fox
Electric Power Research Institute
Palo Alto, CA
USA

ISBN 978-3-319-28951-9     ISBN 978-3-319-28953-3  (eBook)
DOI 10.1007/978-3-319-28953-3

Library of Congress Control Number: 2015960814

© Springer International Publishing Switzerland 2016
This work is subject to copyright. All rights are reserved by the Publisher, whether the whole or part of the material is concerned, specifically the rights of translation, reprinting, reuse of illustrations, recitation, broadcasting, reproduction on microfilms or in any other physical way, and transmission or information storage and retrieval, electronic adaptation, computer software, or by similar or dissimilar methodology now known or hereafter developed.
The use of general descriptive names, registered names, trademarks, service marks, etc. in this publication does not imply, even in the absence of a specific statement, that such names are exempt from the relevant protective laws and regulations and therefore free for general use.
The publisher, the authors and the editors are safe to assume that the advice and information in this book are believed to be true and accurate at the date of publication. Neither the publisher nor the authors or the editors give a warranty, express or implied, with respect to the material contained herein or for any errors or omissions that may have been made.

Printed on acid-free paper

This Springer imprint is published by SpringerNature
The registered company is Springer International Publishing AG Switzerland

# Foreword

After two-plus decades working to advance sustainable development, I have come to the conclusion that there are only a few sustainability challenges that really matter. This book focuses on one of them: How to transform the energy sector into suppliers of clean, dependable, affordable energy services? *Sustainable Electricity* tackles the thorny issues of what sustainability means in practice, how to reconcile demands from competing stakeholders—consumers, investors, regulators, environmentalists, and others—and how to turn sustainability into a business opportunity.

I've spent most of my career developing tools to help corporate leaders make decisions that are good for people and the planet, and good for their bottom line. Over a decade ago, my colleagues at WRI and I published *Green Ledgers*, providing practical examples of companies using environmental cost information to improve profitability and reduce environmental risk. We followed *Green Ledgers* with *Measuring Up*, which defined a set of corporate environmental performance metrics that were later incorporated into the Global Reporting Initiative framework. I then turned my attention to greenhouse gas emissions, working with the electric power sector and others to develop the Greenhouse Gas Protocol—now the generally accepted international accounting and reporting standard for businesses.

In short, my focus has always been on developing usable tools, not theoretical constructs. It is this focus on the practical that makes *Sustainable Electricity* so compelling to me. It cuts to the chase, compiling practical industry-told case studies of how electric power companies are already moving toward more sustainable electricity.

This book's timing could not be better; the status quo of the North American electric power sector is being challenged on multiple fronts. In 2015 the US Environmental Protection Agency finalized a landmark carbon dioxide regulation to slash emissions from existing power plants. Effluent Guidelines for Steam Electric Power Plants, also finalized in 2015, have imposed limits on water discharges. A historic climate deal was signed in Paris, galvanizing international efforts to limit global warming to 2 °C. And the transport sector has begun its inevitable shift to electric vehicles.

But while change is in the air, we have not come far enough, fast enough. As the world's population expands, urbanizes, and grows wealthier, demand for sustainable energy outstrips supply. More than 1.3 billion people lack access to affordable, reliable electricity, and over 40 % of the world's electricity generation still relies on coal burning. Global renewable energy deployment needs to more than triple by 2020 if we are to meet the 2 °C climate stabilization goal.

When I published *Green Ledgers* many years ago, it coincided with the arrival of my second daughter. Although Serena will soon graduate from college, sustainability has still not graduated from the periphery to the mainstream of business practice. Examples of success remain company-by-company, project-by-project. And many successes have been in sectors that are relatively "easy" to shift. The electric power industry faces formidable sustainability challenges. It is complex, heavily regulated, provides life-supporting products, includes diverse business structures, and is in the midst of a massive transition to "the power system of the future."

Achieving sustainable electricity involves weighing the needs of multiple actors. Making decisions that satisfy regulators, meet shareholders' demands, respond to cost-conscious customers, and balance conflicting demands from environmental groups is no easy task. If these groups' priorities were aligned, company choices would be simple. Yet, as this book notes, sustainability is subject to difficult tradeoffs, among stakeholders and over time. It isn't just about climate change, but about water, species, communities, employees, and the next generation too. It is not surprising that the need to respond effectively to all these perspectives has paralyzed many companies, inhibiting bold decisions that might later be challenged.

What will move us along then? Books like this that get to the heart of the real challenges associated with sustainability. In this book, Jessica Fox and her cohort of authors advance our collective understanding of what it means for an electric power company to be sustainable. I hope you enjoy reading it as much as I did. When you've finished, ask yourself, as a customer, investor, regulator, or activist, how you can act to make this industry more sustainable. Business as usual is no longer an option.

Janet Ranganathan<br>
Vice President, World Resources Institute

**Janet Ranganathan** is the vice president for Science and Research at the World Resources Institute, an action-oriented global research organization that works in more than 50 countries, with offices in the USA, China, India, Brazil, Indonesia, and Europe. She works across WRI's six programs—Food, Forest, Water, Climate, Energy, and Cities—playing a lead role ensuring research is robust and actionable. Ranganathan has written extensively on a broad range of sustainability topics, including business and markets, corporate environmental performance measurement, environmental accounting, climate change, greenhouse gas measurement and reporting, ecosystem degradation, ecosystem services, food security, and global environmental governance.

# Generation Z on Sustainable Electricity

In the sustainability space, adults spend a lot of time talking about future generations. Appeals to ensure the well-being of our kids pull at our deepest emotions. But, how often are we inviting these kids and young adults to the table for discussions? Are we hearing from Generation Z (born starting mid-1990s)? As we work on their behalf, let's make sure we take time to hear directly from our future leaders.

Jessica Fox

## Ronald Burke, Jaiden Fox, Sean Sanders, 7th Graders, San Francisco Bay Area, California

> "Thinking about it, we realize we have grown kind of fond of electricity."

We use electricity so much and get so used to it, and we forget how amazing it is. We are basically trading money for something that is kind of like magic. It is this semi-physical element that some dude discovered by getting his kite zapped by lightning (what was his name … Franklin?). Now all we have to do is plug something in and flash, the light comes on. We forget what had to happen to make this power, move it through wires and poles, arriving like magic right into our warm bedrooms. Thinking about it, we realize we have grown kind of fond of electricity.

We are far from experts, but even we know the importance of electricity. Food and precious resources would be wasted without it because we could not cool or heat them easily. My teachers assign homework and assume that we have access to power. In fact, nearly the entire education system depends on electricity to run classes, do computer research, and even turn in assignments. People's way of life would be completely changed if we didn't have electricity. Every day the sun would set and the whole world would fall into a deep darkness.

We are just 12-year-old middle-school kids and have lots of learning to do, but there are some things we have learned already, simple things like the importance of pollination for our food supply and being really careful about using water—especially in California where we've had a drought since we were 8 years old. We also know that we cannot have all this electricity for nothing in return. To get the land that we need to make power, we might have to destroy animals or insects. Maybe in the future we can figure out how to use crazy fuels such as bacteria, sludge, and vibrations.

One of our little brothers calls power plants "pollution plants" because he always forgets the real name. We're not sure whether our opinion really matters to the companies, but what are we supposed to do as kids if we have to use power but we don't want to impact the ecosystem? If I told my teacher I couldn't do my homework because I was worried about the environment and didn't want to turn on my lights to see my paper, I don't think he would be very impressed. It is pretty extreme to live without electricity and that's why it is so hard. If we could ask the power companies for one thing, we'd really like to be able to do our homework at night and play games on the computer without jeopardizing the environment.

Someday we will buy houses, we will have jobs that will require a lot of electricity, maybe have some kids, but we'll definitely have dogs. We want to live in a sustainable and comfortable way. It might take a long time, we might even be dead before it happens, but the important thing is to keep trying to be sustainable no matter how long it will take. Sustainability is the one elusive issue that humans cannot seem to grasp without some sort of cost, like to the environment or to companies, or to ourselves. We must find sustainable living without a cost—that is the future.

## Julia Rose, 9th Grade, Maryland

> "I'm worried that if we don't go in a better direction with electricity production and use, the world might be in huge trouble."

Electricity is generally a good thing, but it needs to go in a better direction in terms of how it is produced and used. I consider electricity to be very important in my life. There is rarely a moment when I'm not using it, as I depend on it for lighting, heating, air-conditioning, charging my electronic devices, entertainment, and countless other things. I'm aware of the fact that not everyone has electricity, and I believe that having it would help make their lives better by providing hot water, refrigeration, light, and much more.

Although I know that electricity is extremely helpful, there are some adverse side effects to it. I'm concerned that too much dependence on fossil fuels could produce excessive air pollution and carbon dioxide, which would affect the air quality and the climate. And, inefficient use of electricity would make more than would be necessary, which would also result in more air pollution and carbon dioxide.

In the future, I hope for more renewable energy to be used and for less dependence on fossil fuels. Contractors and architects could include solar panels in the construction of buildings, and wind turbines and wave energy transformers could help to harvest renewable energy that is already being made by nature. Thinking more about how much power we use and switching to renewable sources are just a few ideas. In general, I'm worried that if we don't go in a better direction with electricity production and use, the world might be in huge trouble. Climate change might cause the loss of species, as well as coastal areas. The way that ecosystems work would need to evolve and our entire world as we know it would change. I worry about what this change would mean for me, my community, and others.

I'm glad that there are people thinking about better ways to make and use electricity, and that there are books like this being written to help protect our future.

## Chloe Carrow, 18 Years Old, Philadelphia, Pennsylvania

> "Can we develop a society that naturally incorporates 'the right thing' into the way we do business… sustainability by default?"

As an 18 year old in 2016, I look around my college library and almost every student is either on a laptop or smartphone. Each generation is so different and defined by the relevant issues of their time. Past US generations experienced wars, stock market crashes, poverty, shortages, and social revolution. As a Z generation, we have grown up experiencing some of the most damaging natural disasters (like Hurricane Sandy and Katrina, earthquakes, and tsunamis). We have also seen terrorist acts (9–11, school and public shootings, bombings, etc). And we have heard the back and forth debate about climate change ("the fifteen hottest years have been since the year 2000"). Yet, because we grew up with these issues constantly churning in the news, we view them as "our normal."

A vast majority of Americans today grew up where it was *easy* and *fast* to access almost anything. This expectation of instant access to endless supplies of energy, food, water, clean air, and other life necessities has led us to take for granted these essentials. Electricity is a perfect example. A huge factor of our everyday life today is technology. Where ever you go, you are sure to find some sort of electronic device designed to improve convenience. As advanced as we are becoming, we are also increasingly dependent upon electricity—and feel a bit lost without it.

Not many people of any generation think too much about where electricity comes from and assume that the companies that generate it are doing it the best way possible. It's easy to forget about since it's always there exactly when we need it and we don't have to worry about how it got there. How do we become more aware of the affect we are collectively having on our planet? Many of us might think we don't have an impact at all. I remember a social media post: "Imagine if trees gave free wifi. We'd all be planting like crazy. It's a pity they only give us oxygen."—Unknown, 2015

To learn about the struggles that these companies face in trying to do the right thing is amazing. I had no idea how hard it was to bring power to everyone, while making everything run so that we do have power day in and day out. I'm more aware of how I could be changing or doing differently to keep my own life sustainable. Going forward I wonder, specifically for my Z generation, how we will find a way to advance technology while creating cleaner ways to live. Will we bring fresh ideas to rethink what currently exists? Will we be allowed? Can those that set up the systems of the past be open to our ideas of the future? Can we develop a society that naturally incorporates "the right thing" into "the way we do business" … sustainability by default?

The kids of my generation are definitely different than any that has come before. We are about the here and now. We value the fact that the content we consume is readily available at *any* time and there doesn't seem to be anything that isn't possible. The stories set forth in this book begin to lay the groundwork and makes me want to be part of the solution, helping to solve one of the greatest challenges of our time.

# Company and Author Profiles

**Electric Power Research Institute**
The Electric Power Research Institute (EPRI) conducts research, development, and demonstration relating to the generation, delivery, and use of electricity for the benefit of the public. They are an independent, nonprofit organization that brings together scientists and engineers as well as experts from academia and the industry to help address challenges in electricity.

Their work spans nearly every area of electricity generation, delivery, use, management, and environmental responsibility. They provide both short- and long-term solutions in these research areas for the electricity industry, its customers, and society.

**Jessica Fox** is senior program manager at EPRI, leading efforts on water quality trading, ecosystem services, sustainability, and related work. Fox leads EPRI's Energy Sustainability Interest Group, a collaborative forum for electric power company sustainability managers to strategically advance sustainable generation and distribution of electricity. Under her leadership, this effort has grown to include 45 electric utilities and is the largest collaborative group in the industry addressing sustainability. Fox also created and manages the EPRI Ohio River Basin Water Quality Trading Project, which is the world's first interstate trading program for nutrients and won the United States Water Prize in 2015. Under Fox's direction, the project has addressed critical social, scientific, and economic issues and become a model project in the USA (http://wqt.epri.com). Fox is a thought leader on ecosystem services, environmental credit stacking, water quality trading, species conservation banking, and corporate sustainability strategies.

Fox has hosted and facilitated multitudes of stakeholder workshops, authored over 40 reports, and her projects have been covered by Wall Street Journal, The Economist, and other national media outlets. She is a trained meeting facilitator and conflict resolution mediator, a certified ecologist by the United States Ecological Society of America, and a certified habitat steward by the National Wildlife Federation. She interacts with a wide variety of energy and environmental stakeholders, including the US Environmental Protection Agency, the

US Department of Agriculture, the US Fish and Wildlife Services, farmers, and corporate industries. Scientific credibility combined with proven success at teambuilding, conflict resolution, and fund-raising allows her to execute complex, multifaceted projects. Fox received a Bachelor of Science in biological sciences from UC Davis and a Master's degree in biological sciences from Stanford University.

**Anda Ray** is vice president and chief sustainability officer with the Electric Power Research Institute (EPRI), leading Environmental and Health and Safety research, along with global strategy, government affairs, communications, and external relations. Prior to EPRI, Ray spent over 30 years in the electric utility business, primarily with the Tennessee Valley Authority, the largest US Federal Utility and a Resource Development Agency. The breadth of her executive-level electric utility experience includes almost all aspects of the electric industry, including nuclear power, fossil-fueled generation, renewable energy, development of new products and services and emergency response and recovery. Ray's career milestones include establishment of the southeastern US first renewable energy premium program, the executive responsible for TVA's Kingston Ash Spill recovery; testimony before Congress, including on the Water-Energy Nexus, appearing on "World Business Review" with former US Secretary of Defense Caspar Weinberger, "60 Minutes," National Public Radio (NPR), and recognized by the Harvard Business Review for efforts on corporate performance. She served as the designated federal officer for sustainability while at TVA. She also collaborated with the Secretaries of Energy and Agriculture on delivering the first "US Roadmap for Biomass." Her undergraduate degree is in nuclear physics from Auburn University, and an advanced degree in solid state physics from Emory University in Atlanta, Georgia.

**Clark Gellings** is an EPRI Fellow and plays key roles in developing technology strategy and in R&D planning. Gellings joined EPRI in 1982 progressing through a series of technical management and executive positions including seven vice president positions. He was also a chief executive officer of several EPRI subsidiaries. Prior to joining EPRI, he spent 14 years with Public Service Electric and Gas Company. Gellings received distinguished awards from several organizations, including The Illuminating Engineering Society, the Association of Energy Services Professionals, the South African Institute of Electrical Engineers, and CIGRÉ (International Council on Large Electric Systems). He is the 2010 recipient of EnergyBiz Magazine's KITE (Knowledge, Innovation, Technology, Excellence) Lifetime Achievement Award. In April 2013, he received a Lifetime Achievement Award from EPRI. He is currently a member of the Board of the University of Minnesota's Technology Leadership Institute, a member of several National Academy of Sciences committees, and served as a commissioner on the New York Governor's Infrastructure Commission following Superstorm Sandy.

Gellings has a Bachelor of Science in electrical engineering from Newark College of Engineering in New Jersey, a Master of Science in mechanical engineering from New Jersey Institute of Technology, and a Master of Management

Science from the Wesley J. Howe School of Technology Management at Stevens Institute of Technology. Gellings is a registered professional engineer, a life fellow, and a preeminent lecturer in the Institute of Electrical and Electronics Engineers, a Fellow Emeritus in the Illuminating Engineering Society, a member of Sigma Xi, the honor society for scientific research, and an Honorary Member of CIGRÉ and past president of its US National Committee.

## American Electric Power

Headquartered in Columbus, Ohio, American Electric Power (AEP) is one of the nation's largest electric utilities in the USA, serving over 5.3 million regulated utility customers in 11 states and an additional 240,000 through its competitive retail energy unit. Its service territory covers 200,000 square miles in Arkansas, Indiana, Kentucky, Louisiana, Michigan, Ohio, Oklahoma, Tennessee, Texas, Virginia, and West Virginia. AEP owns nearly 32,000 MW of generating capacity in the US AEP also owns the nation's largest electricity transmission system, a more than 40,000-mile network that includes more 765-kV extra-high-voltage transmission lines than all other US transmission systems combined. AEP's transmission system directly or indirectly serves about 10 % of the electricity demand in the Eastern Interconnection, the interconnected transmission system that covers 38 eastern and central US states and eastern Canada, and approximately 11 % of the electricity demand in ERCOT, the transmission system that covers much of Texas.

AEP's Generation and Marketing competitive business segment includes subsidiaries that have nonutility generating assets, a wholesale energy trading and marketing business, and a retail supply and energy management unit.

**Sandy Nessing** with nearly 20 years in the electric and natural gas utility industries, Sandy Nessing leads AEP's sustainability strategy and governance, internal and external stakeholder engagement, and annual performance reporting. AEP was among the first US companies to adopt a comprehensive approach to integrated performance reporting, presenting both financial and nonfinancial data, which her team spearheaded. Internally, Nessing leads collaboration on sustainability issues across AEP, raising awareness of functional, resource, and issue interdependencies. A trained *Master Culture Champion*, Nessing supports AEP's culture journey, providing counsel to senior executives and leading efforts to advance a culture shift within AEP that aligns business strategies with employee engagement and performance management. Externally, she identifies and manages advocacy opportunities and provides guidance on fostering relationships with various stakeholders that lead to collaborative initiatives in support of AEP's business strategy and protects AEP's reputation. She is AEP's primary liaison with environmental organizations.

Prior to joining AEP, Nessing led corporate communications for Yankee Gas Services Company (now Eversource Energy), Connecticut's largest natural gas distribution company. Before her work in the utility industry, she led external communications for a healthcare system and launched her career as a broadcast journalist in Connecticut, holding positions that included radio news director, news anchor, and reporter. She earned her degree as a broadcast specialist from Briarwood College in Southington, Connecticut.

## BC Hydro

BC Hydro is a commercial Crown corporation owned by the Province of British Columbia. BC Hydro is one of North America's leading providers of clean, renewable energy, and the largest electric utility in British Columbia, serving approximately 95 % of the province's population and approximately 4 million customers. BC Hydro generates between 43,000 and 56,000 GWh of electricity per year through a network of over 77,000 km of transmission and distribution lines. Over 90 % of the company's generation is hydroelectric, with the remaining generation coming from thermal and diesel sources. In 2012, the company's revenues were approximately $5.7 billion and they have approximately 5800 employees.

**Brenda Goehring** has 27 years working across BC Hydro, and is currently Senior Manager of Corporate Policy and Reporting. This function provides proactive corporate policy analysis and recommendations, as well as managing the external performance and compliance reporting to regulators and associations, and internally to the Executive Board and Board of Directors. Goehring previously held leadership roles across the organization ranging from corporate sustainability, marketing and sales, corporate communications, community relations, environmental nongovernmental organization liaison, public education, electricity trade, integrated resource planning, and green energy acquisition and development.

She has represented BC Hydro as cochair of the Electric Power Research Institute Energy Sustainability Interest Group, on the leadership council for the Ivey School of Business Network for Business Sustainability, the Canadian Hydropower Association Clean Energy Policy Working Group, the BC Business Council Environment Committee, and as an advisor to the Pacific Impacts Climate Consortium and Natural Resources Canada's Climate Adaptation Energy Sector Working Group. Goehring has served on multiple expert panels providing feedback on corporate social responsibility reports by peer companies and also speaks on sustainability topics at forums hosted by the World Bank International Finance Corp. and the UCLA School of Public Policy.

**Kristin Hanlon** currently leads the Strategic Planning function in BC Hydro's Conservation and Energy Management organization, which is responsible for developing BC Hydro's long-term demand-side management plans, leading regulatory application development as well as managing Power Smart's external stakeholder and First Nations advisory committee. Hanlon has over 15 years of experience in the electricity and gas energy utility business and has worked in a number of areas with direct touchpoints to sustainability, such as green energy development, corporate environment, and long-term resource planning. In particular, Hanlon has led a number of projects involving environmental externalities in decision making. Hanlon has a Bachelor of Arts in economics from Simon Fraser University and has completed graduate studies in economics at the University of Calgary.

## Consolidated Edison Company of New York

Consolidated Edison Company of New York (Con Edison) provides electric service to approximately 3.4 million customers in New York City (except for a small area of Queens), and most of Westchester County. The company provides natural gas service to over one million customers in Manhattan, the Bronx, and parts of Queens and Westchester. Con Edison also owns and operates the world's largest district steam system, providing steam service to approximately 1,700 customers in Manhattan. Con Edison is a subsidiary of Consolidated Edison, Inc., which had 2012 revenues of $12 billion and 14,529 employees. Overall, Con Edison's service territory covers 660 square miles, serving a population of nearly 10 million and more than 50 million annual visitors to New York City.

**Jim Skillman** is a distributed generation project manager for the Consolidated Edison Company of New York. As part of Con Edison's central Distributed Generation Group, Skillman has worked to standardize policies across the five boroughs of New York City and Westchester County, improve internal procedures to streamline distributed generation interconnection, and provide customers and installers a central voice for oversight and advocacy for their projects. In his current assignment of distributed generation project manager, Skillman is the photovoltaic process coordinator and project leader for the Smart Grid Photovoltaic (PV) Pilot Program, which recently interconnected a 1.6-MW PV system on an isolated network installation—a first in the northeast. He is the primary point of contact for NYSERDA's Competitive PV and Combined Heat and Power programs, is a member of the City University of New York's Permitting and Interconnection Working Group, and is a member of the NYSERDA Low to Median Income and Community Solar Working Groups. Skillman has over a decade of utility experience, with half of that time working directly with distributed generation installers. He joined Con Edison in 2006 as an Electric Operations Field Supervisor for the Equipment Group, leading field crews in the maintenance, repair, and replacement of low-voltage transformers and network protectors throughout the Brooklyn/Queens territory. As a member of Brooklyn Energy Services, he personally inspected and approved 35 distributed generation installations totaling over 2 MW of capacity. Following Superstorm Sandy, he served as the primary point of contact for the restoration efforts at the Coney Island Aquarium, Kingsborough Community College, and the New York City Parks Department reconstruction of the Coney Island Boardwalk.

Skillman received a Bachelor of Science in chemistry from the US Naval Academy and served in the Navy as a Nuclear Surface Warfare Officer. He also holds a Master of Science in organizational leadership from Mercy College and has earned a project management professional certificate.

**Morgan Scott** was the sustainability manager at Consolidated Edison Company of New York from January 2012 to January 2015. In this role, she managed the company's sustainability strategy and associated initiatives, including the redesign of their sustainability strategy to better align it with the triple bottom line concept and the company's material issues. Additionally, she managed the production of the company's annual sustainability report and voluntary reporting to organizations such as CDP and the Global Reporting Initiative, among others. Scott also held responsibility in the areas of procurement, customer service, energy management, and environment, health, and safety. Scott joined the Electric Power Research Institute (EPRI) in January of 2015 as a sustainability technical lead and project manager. She received a Bachelor of Science, summa cum laude, in business administration from Wagener College and a Master of Science in sustainability management from Columbia University.

## CPS Energy

CPS Energy is the nation's largest municipally owned energy utility providing both natural gas and electric service, serving more than 770,000 electric customers and 336,000 natural gas customers. They have a diverse generation mix that includes coal (41 %), nuclear (24 %), natural gas (18 %), renewable energy, including wind, solar, and landfill-generated methane gas (11 %), and purchased power (3 %). Their 2014 revenues were $2.6 billion, and they have approximately 3000 employees.

**Kim Stoker** is the director of environment and sustainability at CPS Energy. Stoker has been part of CPS Energy's Environmental program since 1989 and is responsible for the development of corporate environmental policies, goals, and metrics designed to support the mission of CPS Energy. She represents the interests of CPS Energy on several regional air, water, and sustainability planning committees. Stoker leads a team of managers and technical experts responsible for environmental strategy, permitting and consulting, in meeting compliance requirements and implementing sustainable business practices. She has a Bachelor of Science in geology and a Master of Science in hydrogeology and is a registered environmental manager and a professional geologist in Texas.

**Lisa Clyde** is part of CPS Energy's Environmental and Sustainability department. Her primary role is air quality compliance. Her responsibilities include environmental reporting for CPS Energy's power plants such as the annual Emission Inventory and EPA Greenhouse Gas Mandatory Reporting Rule. Clyde also was part of the team at CPS Energy that developed the company's first Sustainability Report in 2010 and continues to support efforts for sequential reports. Clyde has a Master of Science from the University of Texas at San Antonio and a Bachelor of Science in chemical engineering from Texas A&M University. She has a professional engineering license in the State of Texas and maintains a Leadership in Energy and Environmental Design (LEED) Green Associate accreditation.

Company and Author Profiles                                    xvii

**Monika Maeckle** works as principal at the San Antonio-based communications consulting boutique, The Arsenal Group. With decades of media and marketing experience, most recently as director of integrated communications at CPS Energy, Maeckle brings a broad perspective to her work with topic expertise in conservation, new energy, and new media.

## DTE Energy

DTE Energy Co. develops and manages energy-related businesses and services nationwide. Its largest operating subsidiaries are DTE Electric and DTE Gas. Together, these regulated utility companies provide electric and/or gas services to more than three million residential, business, and industrial customers throughout Michigan. DTE Energy generates, transmits, and distributes electricity to 2.1 million customers in southeastern Michigan. They have 11,084 MW of generation capacity that includes coal, nuclear fuel, natural gas, hydroelectric pumped storage, and renewable sources. DTE Gas purchases, stores, transmits, distributes, and sells natural gas to approximately 1.2 million customers in Michigan. The company owns and operates 278 storage wells representing approximately 34 % of the underground working capacity in Michigan. There is more gas storage capacity in Michigan than in any other state. DTE Energy's 2012 revenues were $8.8 billion.

**Patricia Ireland** leads DTE Energy's Pollution Prevention and Clean Corporate Citizen programs, establishes and tracks sustainability metrics, and plays a key technical role on DTE's Corporate Citizenship Report team. Ireland joined DTE in 1995 and made a career of developing and implementing programs "beyond compliance," including ISO 14001 certifications (substations, peaking units, electric power plants, gas compressor stations, storage fields, and gas distribution), a vendor environmental liability audit program, green team initiatives, and wildlife habitat certifications. She has also led compliance and cost efficiency-related work such as a vault dewatering strategy, leading the continuous emissions monitoring taskforce, and ozone depleting substance management. Ireland also authored DTE's Environmental Policy and introduced company-wide environmental programs (as a mechanism to ensure environmental consistency) and served for three years on the Chairman's ethics council. Ireland has a Bachelor of Science from Miami University and a Master of Science from Michigan State University in environmental engineering.

## Duke Energy

Duke Energy is the largest electric power holding company in the USA, with approximately $121 billion in total assets. Its regulated utility operations serve approximately 7.3 million electric customers located in six states in the Southeast and Midwest. Its commercial power and international energy business segments own and operate diverse power generation assets in North America and Latin America, including a growing portfolio of renewable energy assets in the USA. Headquartered in Charlotte, North Carolina, Duke Energy is a Fortune 250 company traded on the New York Stock Exchange under the symbol DUK.

**Michelle Abbott** is sustainability director for Duke Energy where she plays a key role in the development and execution of Duke Energy's strategy to operate in a way that is good for people, the planet, and profits. This includes leading the company's internal employee engagement program and representing the company externally to key industry groups. She has also managed the production of the company's annual sustainability report, corporate sustainability goals, and benchmarking efforts. Abbott joined Duke Energy in 1986 as an associate engineer. After a series of promotions within the power delivery department, she was named to senior positions in strategic planning; environment, health, and safety; external relations; and marketing. A native of Raleigh, N.C., Abbott earned a Bachelor of Science, magna cum laude, in industrial engineering from North Carolina State University. She is a registered professional engineer in North Carolina and South Carolina.

## Entergy

Entergy corporation is an integrated energy company engaged primarily in electric power production and retail distribution operations. Entergy owns and operates power plants with approximately 30,000 MW of electric generating capacity, including nearly 10,000 MW of nuclear power, making it one of the nation's leading nuclear generators. Entergy delivers electricity to 2.8 million utility customers in Arkansas, Louisiana, Mississippi, and Texas. Entergy has annual revenues of more than $12 billion and approximately 13,000 employees.

**Brent Dorsey** is a utility veteran with over thirty-five years of broad energy and utility industry experience with a proven track record building and leading successful teams and driving results in both operations management and corporate executive assignments. His most recent experience includes a leadership role in building the Safety and Environment organizations at Entergy Corporation and implementing corporate-wide strategies and changes programs to achieve world-class performance for the organization. As Director of Corporate Environmental Programs for Entergy, he was responsible for the development and stewardship of top-level corporate environmental policies and programs related to adaptation, resiliency, climate change, governance, sustainability and regulatory policy. Entergy has been selected by the Dow Jones Sustainability Index for the past twelve years and named "Best-in-Class" by Carbon Disclosure Project for the past nine years. Dorsey represented Entergy as a member of the stakeholder panel on the Regional Greenhouse Gas Initiative (RGGI) in the Northeast USA. During his career, Dorsey's accomplishments included the acquisition of six merchant nuclear power plants, merchant project development, power marketing and trading, and various planning roles.

Dorsey received a Bachelor of Science in accounting from Louisiana State University, Baton Rouge, LA, and an MBA from Lamar University, Beaumont, TX. Dorsey is a member of the Emissions Marketing Association, Beta Gamma Sigma (honorary business), and Sigma Iota Epsilon (honorary management).

## Exelon Corporation

Exelon Corporation is the nation's leading competitive energy provider, with approximately $25 billion in annual revenues. Headquartered in Chicago, Exelon operates in 48 states, the District of Columbia and Canada. Exelon is the largest competitive US power generator, with approximately 35,000 MW of owned capacity comprising one of the nation's cleanest generation fleets. The company's Constellation business unit provides energy products and services to approximately 100,000 business and public sector customers and approximately 1 million residential customers. Exelon's utilities deliver retail electricity to more than 6.7 million customers in central Maryland (BGE), northern Illinois (ComEd), and southeastern Pennsylvania (PECO) and natural gas to 1.2 million PECO and BGE customers. Exelon has been a leader in providing low-carbon generation since its formation in 2001. Exelon has consistently advocated for a price on carbon and vocalized the importance of climate change action. In 2008, Exelon created its cornerstone Exelon 2020 strategy pledging to abate 15.7 million metric tons of GHG emissions in a year by 2020. Focusing its own internal actions, as well as those of its customers and how it could impact overall grid emissions, Exelon 2020 pushed the bounds of traditional GHG accounting to use a value chain perspective. Despite the failure of timely federal legislation on carbon, a depressed national economy and a sizable merger with Constellation, Exelon was able to achieve its goal seven years ahead of schedule. Read further about how this program helped to shape the company, supported business decisions, and prepared and positioned the company for success in the future as the climate change story continues to unfold and evolve electricity markets.

**Chris Gould** in his role as Senior Vice President, corporate strategy and chief sustainability officer, Chris Gould is responsible for overall corporate strategic planning as well as the company's corporate environmental efforts. He has been with the company since 1999, having previously served as vice president of Corporate Strategy and Exelon 2020. Previously Gould served as vice president, corporate planning where he was responsible for Exelon's planning and project evaluation functions, as director of pricing and structuring, as director of financial planning and analysis, and as manager of market planning at Exelon Power Team, located in Kennett Square, PA. Before joining Exelon, Gould was with Dames and Moore (URS Corporation) in Washington, D.C., serving in a variety of engineering and project management roles in the environmental and infrastructure professional services sector. He began his career at EA Engineering as a project engineer working on various federal, state, and local government agency and public utility projects to protect and remediate the environment. He earned a Bachelor's degree in civil engineering from the Pennsylvania State University in University Park, PA. He received his MBA in finance from the University of Pittsburgh.

**Bill Brady** leads Exelon's governance and oversight functions for environmental and safety management, which includes the Exelon generation, utility and the Constellation energy services companies. His organization supports the senior leadership team with the development and execution of Exelon's sustainable business and environment strategies. His team also is responsible for public environmental reporting, including the annual Exelon Corporate Sustainability Report, the Dow Jones Sustainability Index, and the Carbon Disclosure Project questionnaires. In 2014, he assumed responsibility for implementing a corporate-wide program for auditing environmental and safety compliance. Brady has a Master of Science in mechanical engineering from Drexel University and a Bachelor of Science in mechanical engineering from the University of Rhode Island. He has also completed the Kellogg Executive Development Program at Northwestern University.

**Bruce Alexander** managed the development of Exelon's first corporate-wide greenhouse gas emission (GHG) goal that was negotiated with the US EPA and established under the Agency's Climate Leaders Partnership program in 2003. He also established and managed Exelon's first 3rd party certification of the Exelon Corporation GHG inventory for the period 2001–2008 to a reasonable assurance level, in accordance with the US EPA Climate Leaders GHG Inventory Protocol and ISO 14064:1. He is now responsible for the development and publishing of the annual Exelon Corporate Sustainability Report, as well as maintaining the Exelon Environmental Management System and the emerging legal and regulatory requirements review process.

**Melanie Dickersbach's** responsibilities include managing the greenhouse gas inventory for the corporation, as well as administrating the corporation's Exelon 2020 program designed to cost effectively abate 17.5 million metric tons of GHG emissions by 2020. Dickersbach also supports the evolution of broader sustainability strategies within the corporation, as well as internal and external annual corporate sustainability reporting and web communications. Prior to joining Exelon, Dickersbach worked for a variety of industries providing environmental permitting and compliance support. Dickersbach has a Bachelor of Science in international environmental studies from Rutgers University and a MS in environmental policy from the New Jersey Institute of Technology.

**Alfred Picardi** is a member of Exelon's Corporate Environmental Strategy Team and is currently tasked with developing water and land sustainability programs, metrics, and goals in a consultative approach with the business units, as well as communicating emerging environmental issues and technologies across the company. As manager of Constellation Energy Environmental Affairs and Oversight, he provided analyses and developed technical comments on proposed legislation and regulations in the energy sector. As an independent consultant for World Bank and other multilateral institutions, he worked on international energy and infrastructure development projects, providing environmental impact and rapid strategic screening assessments, environmental management reviews for institutional strengthening, environmental program development and preparation of loan appraisal documents, as well as project implementation oversight. His project portfolio included major power generation, oil pipelines and refineries, as well as renewable energy projects throughout Asia; hydropower, highway, and flood management infrastructure in Vietnam and Indonesia; and health sector programs in the South Pacific. Beginning his career as a consultant in the USA for the environmental impairment liability insurance industry, he performed risk assessments and loss control surveys in a wide range of industrial categories.

## Hoosier Energy Rural Electric Cooperative

Hoosier Energy Rural Electric Cooperative (Hoosier Energy) is a generation and transmission cooperative providing wholesale electric power and services to 18 member distribution cooperatives in central and southern Indiana and southeastern Illinois. Based in Bloomington, Indiana, Hoosier Energy has over 2400 MW of generating capacity from coal, natural gas, and renewable energy power plants and delivers power through a 1500-mile transmission network. Hoosier Energy's 2013 revenues were $668 million, and it has 485 employees.

**Michalene Reilly** has 26 years of electrical utility experience with an educational background in environmental policy analysis and environmental science. She is a certified hazardous materials manager (CHMM) and in late 2014 semi-retired into the position of manager of Environmental Special Projects. She has a Master of Public Administration in environmental policy analysis from Indiana University and a Bachelor of Arts in political science from the University of Wisconsin, Milwaukee. Reilly has been the driving force behind Hoosier's environmental stewardship program, particularly the education programs.

## Los Angeles Department of Water and Power

Los Angeles Department of Water and Power (LADWP) is the largest municipal utility in the nation and the third largest utility in California. LADWP serves 679,000 water customers and 1.5 million electric customers. LADWP has over 7729 MW of generation capacity from a diverse mix of energy sources including renewable energy (23 %), natural gas (17 %), nuclear (10 %), large hydroelectric (4 %), coal (42 %), and unspecified sources (4 %). The first California utility to reach 20 % renewable energy in 2010, LADWP is on track to meet the state-mandated requirement of 33 % renewables by 2020. LADWP is committed to obtaining 15 % of projected power needs in 2020 from energy efficiency, which means that almost half of the electric supply in the city of Los Angeles in 2020 will come from renewable resources. Increasing renewable energy, continued rebuilding of coastal generation units, replacement of coal, infrastructure reliability investments, ramping up energy efficiency, and other demand-side programs are critical and concurrent strategies being undertaken by LADWP. In 2014, electric operating revenues were $3.3 billion. LADWP has approximately 8800 employees.

**Maria Sison-Roces** is a 25-year veteran at the LADWP, where she serves as a utility administrator in the Office of Sustainability and Economic Development, Environmental Affairs. Sison-Roces is responsible for implementing corporate sustainability initiatives and developing strategies to elevate employee awareness on these initiatives. She works collaboratively with the Water, Power, and Joint Systems to manage integrated sustainability reporting for LADWP facilities, including implementing environmentally preferable purchasing and overseeing the LADWP Green Team. In 2015, Sison-Roces led cross-functional efforts to achieve Leadership in Energy and Environmental Design Existing Building Operations and Maintenance (LEED EBOM) Certification for LADWP's headquarters in downtown Los Angeles. From 2008 to 2010, Sison-Roces managed the LADWP strategic planning and change management initiatives while reporting to the Chief Operating Officer. She has a Bachelor of Arts in European languages from the University of the Philippines, a Master's degree in business administration from the University of La Verne, and a sustainability certificate from the University of California, Los Angeles. Sison-Roces currently serves as co-chair for the Publicly Owned Utilities Working Group for the Electric Power Research Institute's Energy Sustainability Interest Group.

**David Jacot P.E.** oversees all aspects of LADWP's offerings and strategies designed to overcome market barriers to the comprehensive adoption of energy efficiency by LADWP's customers, as well as the implementation of LADWP's class-leading water conservation and efficiency programs. Jacot also oversees the integration of water and energy efficiency program delivery across LADWP's service territory as well as through a nation-leading joint program partnership with the natural gas utility serving Los Angeles, the Southern California Gas Company. Jacot has a Bachelor's Degree in Mechanical Engineering from the University of Oklahoma and a Master's Degree in Urban and Regional Planning from California State Polytechnic University—Pomona, as well as 15 years of experience in designing high-performance building systems, modeling building energy usage, and managing cost-effective and investment-grade energy efficiency programs.

## Southern California Edison

Southern California Edison (SCE), a subsidiary of Edison International, is one of the largest electric utilities in the USA and a longtime leader in renewable energy and energy efficiency. With headquarters in Rosemead, California, SCE serves more than 14 million people in a 50,000 square-mile area of central, coastal, and southern California. SCE has provided electric service in the region for more than 125 years. SCE generates about 16 % of the electricity it provides to customers, with the remaining 84 % purchased from Independent Power Producers. SCE's 2014 revenues were approximately $13.4 billion and it has approximately 13,600 employees.

**Dawn Wilson** is the director of Environmental Policy and Affairs at SCE, where she is responsible for Environmental Affairs, Environmental Policy, and Infrastructure Licensing. Wilson's team manages SCE's Corporate Responsibility Projects and Reporting; Environmental NGO Engagement and Grant Requests; Environmental (Air, Climate, Water, Natural Resources, and Public Lands) Policy and Engagement with Government and Regulatory Agencies; and Regulatory Case Management for Generation, Transmission, and Substation Projects.

Wilson joined SCE as an Environmental Compliance Attorney 24 years ago. Prior to her current position, she held roles in a number of areas of the company, including the Law Department and Operational Services. Wilson received a Juris Doctor degree from the University of Michigan and a Bachelor of Arts in political science from Wellesley College.

**Ron Gales** is a senior communications project manager at SCE. His responsibilities include leading the development of SCE's annual Corporate Responsibility Report. A graduate of UCLA, he has been working in Corporate Communications for more than 20 years.

**Pat Adams** is a principal advisor in Regulatory Affairs at SCE, responsible for regulatory strategy and agency interface for infrastructure licensing projects. In her previous role, she served as a senior project manager in Environmental Policy and Affairs where she was responsible for sustainability initiatives for the company, working with internal and external stakeholders to address environmental mitigation and policy issues. Adams has more than 20 years of experience managing projects in the utility industry. She received both an MBA and a Bachelor of Science in psychology from Louisiana State University, in addition to a Project Management Certificate from UC Irvine.

## Tennessee Valley Authority

The Tennessee Valley Authority is a corporate agency of the USA and the largest public power system in the country. It provides electricity for business customers and local power distributors, serving 9 million people in parts of seven southeastern states. TVA receives no taxpayer funding, deriving virtually all of its revenues from sales of electricity. In addition to operating and investing its revenues in its electric system, TVA provides flood control, navigation, and land management (80,000 square miles of public lands, 12,000 archeological sites, and 11,000 miles of shoreline) for the Tennessee River system and assists local power companies and state and local governments with economic development and job creation.

**Tiffany Foster** is the partnerships and educational outreach specialist for TVA's Natural Resources group. She has been with TVA for 13 years. Her work focuses on engaging community members and enhancing awareness and appreciation of the cultural and natural resources of the Tennessee River watershed and the power service area supported by TVA. She earned a Bachelor of Science in Biology with a minor in Chemistry from University of North Carolina Wilmington and a Master of Science in Soil Science and Environmental Sciences with a concentration in water chemistry from the University of Tennessee Knoxville.

**Monte Lee Matthews** is a high-performance leader with over twenty years of diverse experience at the Tennessee Valley Authority, including leadership positions in engineering, energy management, total quality management, and business operations. Lee currently leads TVA's governance and oversight functions for sustainability and climate. He provides counsel relative to the development and execution of TVA's sustainable business and climate strategies. His team also is responsible for customer engagement, key disclosure indices, and public reporting including the annual Strategic Sustainability Performance Plan. Prior to joining TVA, Lee was an architect with CRS-Sirrine, and an energy program manager for DOD-FORSCOM. Lee has a Master's degree in architecture from North Carolina State University, as well as a B.A. in psychology from UNC Chapel Hill, and a B.A. in architecture from UNC Charlotte. Lee has also completed the Kellogg Executive Development Program at Northwestern University, as well as the Wharton School Executive Development Program at The University of Pennsylvania. Lee has authored *Knowledge Driven Profit Improvement*, has been a board regent for NAEM and a member of the EPRI Utility Environmental Benchmarking Forum steering committee.

**David Matthews** is an aquatic zoologist for TVA. He has been involved with stream ecological health monitoring, using Index of Biotic Integrity protocol in the Tennessee Valley by supporting TVA stream monitoring crews as project lead and coordinating cooperative monitoring and support with other agencies. Other duties include IBI database administration and endangered species surveys. Matthews is a member of the Southeastern Lake Sturgeon Working Group. These efforts improve quality of life in the Tennessee Valley by providing resource managers with data and support to make sound environmental decisions. He has been with TVA for 26 years and has an associate in Applied Science in Fish and Wildlife Management from Haywood Community College.

**Hill Henry** leads TVA's Natural Resource Policy in TVA's Environment and Energy Policy group. As a biologist at TVA for 20 years, he has studied endangered species, avian resources, cave ecosystems, and natural resource management in the Tennessee River Valley. In his current role, he works with interagency teams to develop implementation strategies for wetlands, fisheries, NEPA, and cultural and natural resource policies. Hill is a certified wildlife biologist and project manager professional with a B.S. in wildlife sciences and M.S. in zoology from Auburn University.

**Shannon O'Quinn** is a water resource specialist for TVA. His focus is to improve and protect water quality and biodiversity in the Tennessee Valley which covers portions of seven states: Virginia, Tennessee, North Carolina, Georgia, Alabama, Mississippi and Kentucky. This is accomplished by partnering with community groups and resource agencies to assess watershed conditions, develop improvement plans, leverage funding, and implement water quality improvement projects. These efforts positively impact local communities by improving drinking water supplies, aquatic habitat, recreational opportunities, human health, and economic development. O'Quinn has been with TVA for 15 years and has a Bachelor's degree in environmental studies from Radford University and a Master's degree in geosciences from ETSU.

# Preface

This is the only compilation of first-hand industry accounts describing how power companies are making strides toward achieving sustainable electricity in North America. These case studies highlight game-changing efforts, tell candid stories about challenges and process, and forecast the next decade of innovation. This book offers rare insights into the steps, process, and business case for advancing sustainable electricity in North America.

Through candid stories and rare CEO quotes, readers will obtain a window into how companies are balancing conflicting stakeholder demands to become leaders in water management, renewable energy, employee and customer engagement, climate resiliency, and other pressing sustainability challenges. This will provide a blue book for other companies who may want to follow in their footsteps and will expedite stakeholder understanding of realities and opportunities.

Power company managers seeking to identify specific opportunities and understand the process for advancing projects in their own organizations will greatly benefit from the insights shared in this volume. Customers who want to reduce the impacts of their power use will find pathways to improvement, and regulators will gain an invaluable understanding of the steps and costs required to realize the revolutionary changes anticipated in the electric power industry.

I am grateful to the authors who spent more than a year developing this unique resource, to Janet Ranganathan for her insightful Foreword, and to our five youth authors for the non-traditional Generation Z Foreword sharing their raw thoughts on what Sustainable Electricity means. This book will provide a foundation for a new wave of collaboration and innovation.

Jessica Fox

# Contents

1. Introduction: Defining Sustainable Electricity .................. 1
   Jessica Fox and Anda Ray
2. American Electric Power: Stakeholder Engagement and Company Culture ....................................... 7
   Sandy Nessing
3. BC Hydro: How Demand-Side Management Can Help Utilities Achieve Triple Bottom Line Outcomes ....................... 31
   Brenda Goehring and Kristin Hanlon
4. Consolidated Edison Company of New York: Distributed Generation ................................................. 53
   Jim Skillman and Morgan Scott
5. CPS Energy: Clean Energy, Community, and Business Vitality .... 75
   Monika Maeckle, Lisa Clyde and Kim Stoker
6. DTE Energy: Water Management ........................... 91
   Patricia Ireland
7. Duke Energy: Engaging Employees in Sustainability ............ 111
   Michelle Abbott
8. Entergy: Climate Change Resiliency and Adaptation ............ 133
   Brent Dorsey
9. Exelon Corporation: Strategic Greenhouse Gas Management ...... 151
   Christopher D. Gould, William J. Brady, Melanie Dickersbach, Bruce Alexander and Alfred Picardi
10. Hoosier Energy Rural Electric Cooperative, Inc.: The Rural Cooperative Perspective ..................................... 169
    Michalene Reilly

11  **Los Angeles Department of Water and Power: Energy Efficiency for Our City and Our Customers** .............................. 191
    Maria Sison-Roces and David Jacot

12  **Southern California Edison: Renewable Energy** ................. 211
    Dawn Wilson, Ron Gales and Pat Adams

13  **Tennessee Valley Authority: Balancing Aquatic Biodiversity, River Management, and Power Generation** ..................... 225
    Tiffany Foster, Monte Lee Matthews, David Matthews, Hill Henry and Shannon O'Quinn

14  **The Fourth Energy Wave: The Sustainable Consumer Is Knocking** ................................................. 241
    Clark W. Gellings

15  **The Next Decade of Sustainability Science** ..................... 257
    Jessica Fox

**Index** ................................................................ 263

# Glossary

**Advanced Metering Infrastructure (AMI)** An integrated system of smart meters, communications networks, and data management systems that enables two-way communication between utilities and customers.

**Biomonitoring** Examining the biological responses of fish and benthic communities to assess changes in the aquatic environment.

**California Renewables Portfolio Standard (RPS)** Established in 2002 under Senate Bill 1078, accelerated in 2006 under Senate Bill 107 and expanded in 2011 under Senate Bill 2, California's RPS is one of the most ambitious renewable energy standards in the country. The program requires investor-owned utilities (IOUs), electric service providers, and community choice aggregators to increase procurement from eligible renewable energy resources to 33 % of total procurement by 2020.

**$CO_2e$** Carbon dioxide equivalency ($CO_2e$) converts a mix of greenhouse gases into the equivalent amount of $CO_2$ that would have the same global warming potential.

**Contract Demand** A fixed charge portion of standby rates that is dependent upon the amount of instantaneous backup power required by the customer upon loss of the distributed generation facility.

**CSR** Corporate Social Responsibility.

**Energy Demand** The aggregate kilowatt-hours (kWh) of energy usage in a given time period expressed as a kilowatt (kW) value. It is used to appropriately size service capacity to customer usage. For example, a customer that uses 100 kW in an hour could be said to have required 100 kWh of demand.

**Demand Response** A program utilized by a distribution or transmission utility to reduce the peak power consumption (demand) on the system for a specified duration and time.

**Distributed Energy Resources (DER)** Smaller power sources that can be aggregated to provide power necessary to meet regular demand.

**Distributed Generation (DG)** An approach that employs small-scale technologies to produce electricity close to the end users of power. Such technologies often consist of modular (and sometimes renewable energy) generators.

**Ecological Conditions Assessment** Monitoring of ecological resources to discover current and changing conditions.

**Economic Dispatch** The short-term determination of the optimal output of a number of electricity generation facilities to meet the system load at the lowest possible cost, subject to transmission and operational constraints.

**Empowerment Zone** An area designated by a local utility within which developers would receive additional incentives to install distributed energy facilities, typically used to serve a utility load requirement or to offset a large capital expenditure by the distribution or transmission utility.

**Global Reporting Initiative (GRI)** GRI is an international independent organization that helps businesses, governments, and other organizations understand and communicate the impact of business on critical sustainability issues such as climate change, human rights, corruption, and many others.

**High Tension Service** Electric service provided to the customer at the higher voltage level of the distribution feeders, typically for larger load applications.

**Home Area Network (HAN)** A type of local area network with the purpose to facilitate communication among digital devices present inside or within the close vicinity of a home.

**Hydrologic Units** A hierarchical classification system developed by USGS to delineate and organize watersheds.

**Index of Biotic Integrity** Index is based on 12 ecological metrics encompassing species richness and composition, trophic structure, fish abundance, and fish condition. It is used to classify the condition of a stream along a continuum of very poor to excellent.

**Interconnection** When a source of local electric generation is operating simultaneously and in parallel with the local utility power source.

**Islanding** A condition where a local distributed generation power source is providing power to nearby loads upon loss of grid power.

**Isolated Network** A utility design where a large load customer is served by a dedicated set of transformers tied together on the secondary voltage to provide redundancy and increased capacity over a direct grid-connected service lateral.

**Isolated Operation** A condition where a facility breaks connection with the electric utility and operates solely on local generation and also known as *islanding*

**Light Emitting Diode (LED)** A light source with benefits over incandescent light bulbs, including lower energy consumption, longer lifetime, improved physical robustness, smaller size, and faster switching.

**Leadership in Energy and Environmental Design (LEED)** A green building certification program that recognizes best-in-class building strategies and practices

**Low-Carbon Generation** Generation from processes or technologies that produce power with substantially lower amounts of carbon dioxide emissions than conventional fossil fuel power generation. It includes low-carbon generation sources such as wind power, solar power, hydropower, and nuclear power.

**Low Tension Service** Electric service provided to a customer at the lower voltage level common to the secondary side of all distribution transformers in the area, typical voltage level for most electric customers.

**Net Metering** A program where the excess on-site generation is allowed to flow out to the utility grid and the meter reverses direction, giving the customer retail credit for exported power.

**Network Protectors** Utility-grade switches equipped with sensing relays that protect the transformer and high-voltage feeder from reverse power flow from the network grid during fault conditions.

**No-Carbon Generation** Generation from processes or technologies that produce power without emitting carbon dioxide *or other greenhouse gases.*

**N-Type Solar Panels** A solar cell is made of two types of semiconductors, called p-type and n-type silicon. The n-type silicon is made by including atoms that have one more electron in their outer level than does silicon. A solar cell consists of a layer of p-type silicon placed next to a layer of n-type silicon.

**Off-Grid** Customers with no electric service who derive all their power needs from on-site sources.

**Power Purchase Agreement (PPA)** A contract between two parties, one who generates electricity for the purpose (the seller) and one who is looking to purchase electricity (the buyer).

**Pulsing** Releasing water through dams at regular intervals throughout the day to help create an essentially steady flow of water within a few miles, maintaining a more constant wetted habitat downstream of the dams.

**Revenue Decoupling** A ratemaking construct where the aggregate delivery of kilowatt-hours through the utility's distribution system is not directly tied to the Public Service Commission-approved utility revenue requirement. This construct allows for distributed generation systems, which provide a portion of the customer's kWh need that the utility would have otherwise provided, to have a smaller impact on the utility's financials.

**Secondary Spill Containment** The containment of hazardous liquids in order to prevent pollution of soil and water. Common techniques include the use of spill berms to contain oil-filled equipment, fuel tanks, truck washing decks, or any other places or items that may leak hazardous liquids.

**Sentinel Aquatic Species** Aquatic animals used to detect changes in the aquatic environment.

**Smart Grid** A system that includes a variety of operational and energy measures including smart meters, smart appliances, renewable energy resources, and energy efficiency resources. Electronic power conditioning and control of the production and distribution of electricity are important aspects of the smart grid.

**Smart Meters** An electronic device that records consumption of electric energy in intervals of an hour or less and communicates that information at least daily back to the utility for monitoring and billing.

**Species of Concern** Species that are declining or appear to be in need of concentrated conservation actions.

**Standby Service Rates** A rate-making construct which attempts to recover local utility stranded asset costs that are no longer recuperated through kilowatt-hour metering charges following the installation of a large distributed generation facility.

**Stream Impairment** A stream with chronic or recurring monitored violations of the applicable numeric and/or narrative water quality criteria.

**Supervisory Control and Data Acquisition (SCADA) System** A set of sensors, relays, and communications equipment that allows for remote operation of utility assets, as well as real-time monitoring of system conditions.

**UL1741** Underwriter's Laboratories test procedure that verifies inverters meet the protection requirements of IEEE Standard 1547.

# Chapter 1
# Introduction: Defining Sustainable Electricity

Jessica Fox and Anda Ray

**Abstract** There are many definitions of sustainability and company examples of how to meet social, environmental, and economic goals simultaneously. However, how can such approaches apply to the electric power industry, which is subject to heavy regulation and unable to forgo safe, affordable, and constantly reliable power? In *Sustainable Electricity* companies discuss some of the most hotly debated challenges of today: renewable energy, water use, species impacts, employee engagement, stakeholder communication, resiliency and climate change, distributed energy, energy efficiency, greenhouse gas emissions, consumer preferences, and business vitality. Unless we are satisfied with vague business commitments, the current definitions of sustainability to protect future generations are too nebulous to set goals, measure progress, and inform balanced outcomes in the electric power industry. This book provides concrete examples of what "sustainable electricity" will encompass in the next decade.

The idea of sustainability is simple: providing what is needed to ensure someone or something remains viable and productive, or ensure they are *sustained*. This simple definition, though accurate, does not reveal the true complexity of its acheivement, especially since the number of "someones" and "somethings" that need to be simultaneously sustained has grown. The recognition of our interconnectedness has forced contemplation not only for the web of life, but also for the integrated web of people, nature, and economics.

---

J. Fox (✉) · A. Ray
Electric Power Research Institute, Palo Alto, USA
e-mail: jfox@epri.com

© Springer International Publishing Switzerland 2016
J. Fox (ed.), *Sustainable Electricity*, DOI 10.1007/978-3-319-28953-3_1

While Nobel Prize-winning economist, Milton Friedman, held that the primary "social responsibility of business is to increase its profits,"[1] the expanded role of companies to consider their social and environmental externalities has become a central point of discussion. "Sustainable" has adopted an environmental focus of protecting natural resources for future generations,[2] and research has increased awareness that continued business success to achieve Friedman's priority will rely on more diligent consideration of natural assets such as food, fiber, and water. Consideration for this "Natural Capital"[3] can not only lead to an "Eco-Advantage"[4] for business, but could also become necessary for sustained business operation. Companies such as Unilever, Patagonia, and Ben & Jerry's have shown that sustaining a business and sustaining the social and natural community are compatible. They are "Doing Well by Doing Good," showing that it is possible to achieve positive "triple bottom line"[5] outcomes. But, what happens when companies cannot achieve success in both business and social goals? What do CEOs do when there are trade-offs, as is the case with the majority of difficult sustainability-associated choices (Chapter 15)?

The stories of Unilever and Patagonia are inspiring and provide needed lessons for others who are embarking on this journey. These companies and others like them, however, enjoy certain flexibilities in moving their sustainability objectives forward. They can charge prices as they deem appropriate for their products, they can select suppliers that align with company goals (even if they are more costly), and they provide a product that has many market alternatives that give consumers flexibility for choosing other options. These flexibilities mean that their approach does not directly translate into complex industries, such as the electric power sector. The electric industry has unique complexity to being sustainable, which does not relieve their responsibility for taking action, but, it is useful to understand the challenges in order to support viable solutions and recognize that different stakeholders have conflicting demands on the companies. Current trade-offs include reducing water use goals while still providing affordable power or managing sustainable supply chains while complying with least-cost procurement criteria. It is not a surprise that confusion abounds when competing demands of stakeholders, shareholders, and regulators hit the desk of the CEO. Indeed, if it was simple to achieve triple bottom line outcomes in electric power, it would already be widespread!

Alas, some of the characteristics that make this difficult for the electric power industry include the following:

---

[1]Friedman [1].

[2]*United Nations General Assembly (March 20, 1987)*. Report of the World Commission on Environment and Development: Our Common Future; Our Common Future, Chap. 2: Towards Sustainable Development; Paragraph 1. United Nations General Assembly. *Accessed October 2015*.

[3]Daily [2], Hawken et al. [3].

[4]Esty and Winston [4].

[5]Savitz [5].

- It is **heavily** regulated for nearly all aspects of the business, including product price, business operations, land use, waste management, and product delivery.
- It supports diverse residential and industrial customers, who do not have viable alternatives for where to get their product (electricity).
- It provides a life-sustaining product that must be delivered continuously, safely, and at reasonable cost.
- It is subject to shareholder resolutions, customer requests, and environmental lawsuits, some of which demand conflicting actions (lower prices for customers, increased profits for shareholders, and costly technologies to protect the environment).
- Electricity generation has local and global footprints, whether it is renewable or fossil-fuel based.
- The industry is responding to larger "power system of the future" changes.

How can "sustainability" apply to such an industry? When people turn on their light switch, they expect lights to come on and at a reasonable price. The reliability of electricity is considered a non-negotiable expectation that must be achieved. In some cases, electric power companies have a legal obligation to provide electricity to their customers and are regulated in terms of the prices they can charge. In addition to reliability and affordability, power companies must meet targets for employee and public safety during the generation, transmission, and distribution of power. Electricity providers must also comply with a variety of environmental regulations related to air, greenhouse gases, waste materials, species and habitat protection, and electromagnetic fields, among others.

Figure 1.1 illustrates the three pillars of sustainability in the context of the electric power industry's core mission to provide safe, reliable, affordable power. Achievement of a long-term balance across all three pillars of sustainability will likely involve innovation and commitments that extend beyond the breadth of current laws and technologies.

As the electric power industry evolves, so must its approach to sustainability. The ability to integrate and manage emerging innovative technologies, comply with new environmental regulations, identify appropriate voluntary measures beyond regulations, and, especially, meet changing customer expectations are critical components of future success. Identifying a prescriptive approach for achieving sustainable electricity is further stymied by the large diversity of power companies. The global electric power industry is comprised of thousands of companies, some that are owned and operated by various government entities, others operating as profitable enterprises, and still others functioning as cooperatives. As such, it is difficult to have a single step-by-step guide that identifies the approach to achieving a sustainable electric business model within one company, let alone the whole industry.

Which sustainability goals are reasonable to pursue as an industry or company? What does a fully sustainable utility look like?

Progress has been hampered by the complexity of identifying *specific actions* that companies should take. Then, actually measuring, tracking, and reporting

**Fig. 1.1** The three pillars of sustainability within the core mandate of the power industry

sustainability performance presents challenges, especially in cases where there is debate regarding the appropriate metrics to use to measure performance (Chap. 15). For example, given that power companies serve anywhere from thousands of customers up to many million, is it acceptable to normalize greenhouse gas emissions based on the number of customers served? These complexities need to be discussed and researched in order to move from nebulous discussions on protecting future generations into specific goals, scorecards, and commitments.

The reality is that power companies may have to look at more "and" options, not "or" options. Is it possible to have lower greenhouse emissions *and* low price per kilowatt-hour (Chap. 9)? Can we run plants that consume no water *and* still provide reliable power (Chap. 6)? Can we increase the price per kilowatt-hour *and* keep overall customer bills steady through more efficient use of the power (Chaps. 3 and 11)? Will it be possible to protect the resiliency of our ecosystems while providing steady base-load power (Chap. 8)? Where does renewable and distributed energy fit into the expectation for secure and reliable power deliveries (Chaps. 4 and 12)? How will the electric power industry respond to meet the demands of the evolving consumer, who no longer wants "electricity," but rather wants sustained "convenience, comfort, control, and choice" (Chap. 14)?

In *Sustainable Electricity*, we provide real-life stories from companies that have moved the dial on answering these questions. It is the first compilation of industry-told case studies on how power companies are making strides toward sustainable electricity, including unique corporate executive quotes. Chapters discuss some of the most hotly debated challenges of today: renewable energy, water use, species impacts, employee engagement, stakeholder communication, resiliency and climate change, distributed energy, energy efficiency, demand-side management, greenhouse gas emissions, consumer preferences, and business vitality. Each chapter shares topic-focused case studies regarding the challenges, key issues, and reality of implementing sustainable electricity in North America. While admittedly only a beginning, we hope the stories contribute to an active discussion on viable options for progressing on the sustainability journey. Collaboration, honest self-awareness, and acknowledgment of the needs of all players are needed, including the CEO, parent and salmon (Chap. 15).

This book was sponsored by EPRI's Energy Sustainability Interest Group, which is the largest collaboration in the electric power industry around sustainability. The group was launched by EPRI in 2008 and currently has 45 members with combined corporate assets of over $1.1 trillion. The focus of the group is to help members address sustainability challenges and opportunities in the electric

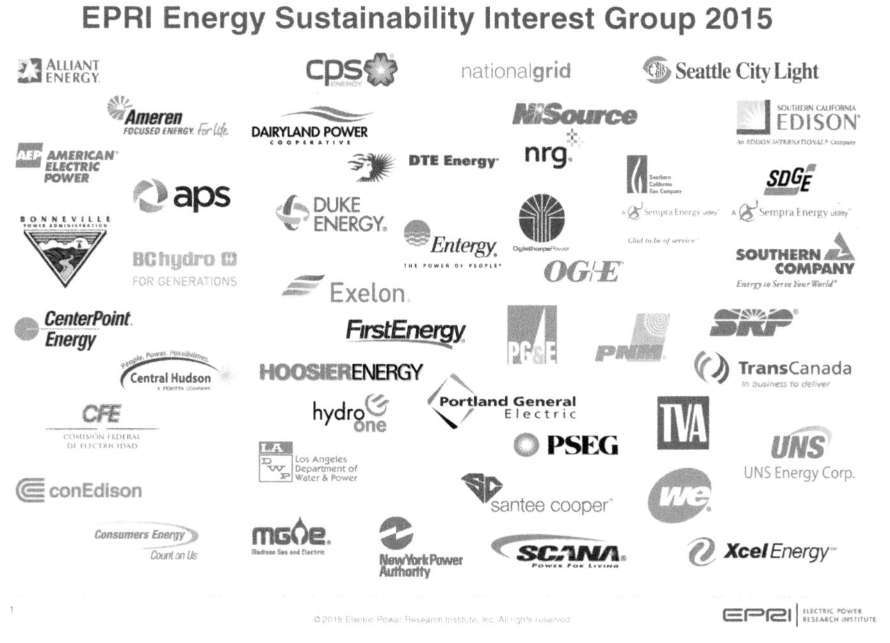

**Fig. 1.2** EPRI Energy Sustainability Interest Group members 2015

industry through collaboration, technical research, and opportunities to interact with key sustainability players.[6] All contributing authors of this book participated in the 2015 Energy Sustainability Interest Group, were not paid for their chapters, were subject to rounds of demanding edits, and were required to provide candid information on their challenges and solutions. We thank them for making this book a reality and providing an important strategic resource to accelerate the adoption of sustainable operating practices in the electric power industry and beyond (Fig. 1.2).

## References

1. Friedman M (1970) The social responsibility of business is to increase its profits. New York Times Mag (Sept 13)
2. Daily G (1997) Nature's services: societal dependence on natural ecosystems. Island Press, Washington, DC
3. Hawken P et al (1999) Natural capitalism: creating the next industrial revolution. Little, Brown and Company, Boston
4. Esty D, Winston A (2006) Green to gold: how smart companies use environmental strategy to innovate, create value, and build competitive advantage. Yale University Press, New Haven
5. Savitz A (2006) The triple bottom line: how today's best-run companies are achieving economic, social, and environmental success—and how you can too. Jossey-Bass, San Francisco

---

[6]Information on EPRI's Energy Sustainability Interest Group can be found at www.epri.com/sustainability.

# Chapter 2
# American Electric Power: Stakeholder Engagement and Company Culture

**Sandy Nessing**

**Abstract** American Electric Power (AEP) is one of the largest utilities in the country with over 18,000 employees. Over the last several years, AEP has been strengthening its corporate culture by applying its approach for stakeholder engagement inward. After an employee culture survey revealed that employees felt disconnected from management decisions and direction, the company's sustainability team collaborated with human resources, business unit leaders, and corporate communications to organize 90 meetings involving more than 1,000 employees to understand why and to explain how the principles of sustainable growth were key to improving AEP's organizational health. These events were a major milestone in AEP's sustainability and culture journeys and marked the intersection of corporate sustainability and employee culture. Sustainability professionals are key to breaking down internal silos and increasing awareness of business strategy and emerging issues or trends for employees and have become the nexus between corporate culture and sustainability.

## 2.1 Introduction

A corporate culture committed to its stakeholders and to sustainable business practices is necessarily based on engagement, transparency, and accessibility. Together these strategies allow companies to better manage risk, build trust, play a role in developing and supporting strong communities, and safeguard the company's

S. Nessing (✉)
Managing Director, Corporate Sustainability, AEP, Columbus, USA
e-mail: smnessing@aep.com

© Springer International Publishing Switzerland 2016
J. Fox (ed.), *Sustainable Electricity*, DOI 10.1007/978-3-319-28953-3_2

financial health. While many companies most often focus this effort on external stakeholders, they overlook the benefits of turning the engagement lens inward. American Electric Power (AEP) has had a formal stakeholder engagement process in place for more than eight years, and its focus was primarily external—until recently.

Stakeholders expect AEP to be transparent, open, and candid. The success of this process became a model for internal engagement when the company sought to improve its internal organizational health. In 2012, AEP applied its approach for stakeholder engagement inward. An employee culture survey was conducted to understand how employees relate to the organization they work for and illuminate the day-to-day culture of AEP. This inward focus signaled to investors and other external stakeholders that AEP's commitment to its own internal culture is equally important as its outside stakeholder meetings, and would be a positive influence on management's performance and the company's strategic direction.

However, as AEP executives reviewed the results of the employee culture survey, they became concerned. While the survey identified strong cultural pillars of safety, commitment to customers, and a solid desire to contribute to AEP's success, an apparent lack of knowledge or understanding of changing business dynamics left employees feeling disconnected. CEO Nick Akins said the survey seemed to indicate that "Management works for one company and employees work for another."[1] The culture of the senior management team was not shared by the 18,500 AEP employees. Akins and other senior leaders understood this gap meant risk and inefficiencies—issues contrary to a sustainable organization. By turning the company's established external stakeholder process inward, executives began to close the gap.

AEP realized its success was tied to understanding the dynamics of employee culture, a tough lesson for companies to embrace; and at the outset, AEP was no different than most. It was very clear to leadership that it needed different inputs to inform decision making, internally and externally. Otherwise, they risked growing resource inefficiency, missing new business opportunities or risks, employee retention and attraction issues, or failing to identify emerging trends. Owing to its positive experience with external stakeholder engagement, the sustainability team that led external stakeholder engagement was now tasked with conducting employee focus groups. The team, working collaboratively with human resources, business unit leaders, and corporate communications, organized 90 meetings involving more than 1,000 employees and gathered a wealth of information and ideas. Before "captive" audiences, the company framed the principles of sustainable growth as necessary to improve AEP's organizational health. This event was a major milestone in AEP's sustainability and culture journeys and marked the intersection of corporate sustainability and employee culture.

---

[1]Personal communication, AEP employee Webcast, August 27, 2012.

This approach appeared to be validated by an investor during an external stakeholder meeting in late 2013 that included environmental groups and investors. Akins mentioned AEP's focus on employee culture and why it was important to him and to AEP's business success. A few stakeholders nodded their heads, while others looked puzzled. One AEP investor said, "A company focused on its culture is a company whose employees understand both the challenges and opportunities, are engaged in moving the company forward and is the hallmark of a company that is well-managed."[2]

Sustainability professionals must work collaboratively across an organization to effectively address stakeholder concerns. They do this by breaking down silos and increasing awareness of business strategy and emerging issues or trends. As a result, management begins to understand the outside world's many different perceptions of AEP and why transparency and engagement are business imperatives. For example, water resource management is an increasingly significant sustainability issue for the electric power sector. Reflecting that issue's importance, AEP reports extensively each year on water withdrawal, consumption, quality, and availability in its Global Reporting Initiative (GRI) report and the Carbon Disclosure Project (CDP) Water Survey.

During preparations for the 2014 Corporate Accountability Report, a power plant engineer asked why the water data were being requested. To his way of thinking, it was not "inside the bounds of doing things legally, ethically, and morally."[3] He said safety, environmental protection, ethics and compliance, and serving customer needs were his primary responsibilities. How did the request for water data fit within those objectives? After learning that the data are used by environmental stakeholders, investors, and others, he was satisfied. "I am a third generation AEP stockholder … as a stockholder, I have been concerned with duplication of effort within the company and the actual necessity of various bits of information going to some of the places it does."[4] With that, he categorized the request under power production (serving customers), thereby matching one of his priorities.

Had that exchange not occurred, the plant engineer would still have provided the data but without understanding its importance or how it was being used. Absent the strategic alignment between the data request and stakeholder questions about how the company manages its water resources, he simply considered it another meaningless, burdensome request.

The ability to connect the dots is one reason why the sustainability professional has become the nexus between culture and sustainability. For as much as external stakeholders want to know why companies do the things they do, employees are equally hungry for that information.

---

[2]Personal communication, AEP stakeholder meeting, November 13, 2013.
[3]Personal communication with AEP employee, March 2014.
[4]Personal communication with AEP employee, March 2014.

## 2.2 How Stakeholder Engagement Began

The path to a formal stakeholder engagement process began in 2004 with a shareholder resolution seeking greater transparency of AEP's air emissions risks. In response, a subcommittee of independent directors of AEP's Board of Directors formed what was essentially a stakeholder group to address the issue. They sought input from the shareholder resolution author and environmental groups, as well as from the internal company experts. When the report was issued later that year—"An Assessment of AEP's Actions to Mitigate the Economic Impacts of Emissions Policies"—it was described by some stakeholders as a groundbreaking example of effective corporate governance and stakeholder engagement. Among the commitments in the report was a pledge to be open to partnerships, in technology and policy, and for continued transparency of actions (Fig. 2.1).

In 2006, AEP chose to move from a biannual report of environmental performance to a broader sustainability report that included its position and approach to climate change, performance of the electric system, and social issues such as the aging work force, worker health and safety, and stakeholder engagement. It would also serve as the vehicle to keep the company's promise of transparency to stakeholders from the 2004 Board report.

One of the first steps was to learn more about the issues important to AEP's stakeholders. Since AEP had not organized a formal meeting with external stakeholders before, it turned to Ceres, a national coalition of investors, companies, and

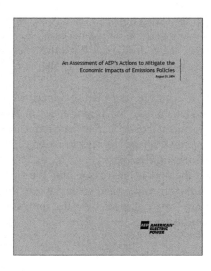

**Fig. 2.1** The 2004 Board Report on Emissions Risk served as the catalyst for greater transparency at AEP

public interest groups that advocates for sustainable business practices, to help convene the first meeting. AEP's then CEO Mike Morris asked his senior team to participate and signaled that stakeholder engagement would become part of how AEP would conduct business going forward.

Ceres worked with AEP to identify the right external stakeholders, the issues to be discussed and the meeting format. Ceres served as facilitator; they had a degree of credibility and trust with the stakeholders that AEP did not yet enjoy. It was a new type of relationship for the stakeholders and for AEP. That first meeting included non-government organizations (NGOs), such as the Natural Resources Defense Council, the Clean Air Task Force, and Environmental Defense Fund. Other participants included labor representatives, investors, and other socially minded groups, such as the Presbyterian Church. The top issues were climate change, environmental performance, future of coal, energy efficiency, and utility business model.

This new experience was initially uncomfortable for many in the room, including stakeholders who did not know what to expect. The dialogue was always respectful and polite, but as each challenged or defended positions, it began to resemble a boxing match. The stakeholders would challenge the company on an issue or question why it held a certain position. AEP would respond, sometimes defensively. Back and forth it went. It was an inevitable outcome since neither side had ever sat face-to-face in such a forum.

AEP and Ceres agreed to call a time out to remind each other of the purpose of the meeting and asking participants to actively listen to each other. That gentle reminder was a turning point. As the conversation picked up again, both groups realized that it was an important first step toward building lasting relationships.

A few days later, the AEP team reassembled to discuss what they heard during that meeting and what actions they would take, if any. Initially, there was frustration that the stakeholders did not know more about or understand AEP's business. It was argued that the stakeholders were asking for things that were already being done (such as investments to improve power plant efficiency).

Almost immediately, one executive stopped the discussion and said that the reason the stakeholders did not know was because AEP had not done a good enough job communicating with them. The group agreed that this was an important opportunity for engaging stakeholders and to tell the AEP story. That was the turning point for stakeholder engagement at AEP; senior executives have been fully committed and have participated in nearly every stakeholder meeting since that first "match."

This demonstrates the degree of value in stakeholder engagement. AEP took it one step further and began convening meetings with NGOs that included customers and regulators. AEP quickly learned that there is no better way to communicate the difficulties of navigating competing interests than by bringing those different viewpoints together. Customers heard NGOs pushing for something, regulators could hear what customers were being asked to pay for, and NGOs heard customers say why they could not afford it and did not want to pay for it.

The multi-stakeholder approach allowed all stakeholders to see the competing demands that AEP balanced and for AEP to learn from many different positions at once. In many cases, these exchanges informed business strategy and led to the creation of new sustainability commitments. For example, during ongoing discussions about advancing energy efficiency with customers, AEP realized it needed to examine its own energy use and set internal energy reduction goals. The company collected a year of data and set goals to reduce consumption. By the end of 2013, AEP had reduced its kilowatt-hour (kWh) usage by 25 % compared to the 2007 baseline. The equivalent accumulated savings from reduced energy consumption at more than 300 facilities exceeded $17 million. The company achieved these results through a combination of capital investments in more energy efficient equipment and an employee education campaign. By reducing usage, the company could sell the unused energy in the wholesale power market, or not produce it at all, while reducing impacts to the environment. One investor who participated in AEP's stakeholder process said it sent a positive signal that "companies do care about their stakeholders and those who impact the long-term value of the company."[5]

Today, AEP's framework for engagement has evolved from large meetings with a broad focus to more targeted, smaller group meetings. Senior executives still participate, including the CEO. The vehicles for engagement have evolved as well. Nothing is more effective and meaningful than face-to-face meetings but that is not always an option. AEP has employed the use of conference calls, e-mail, and social media to stay connected with its stakeholders.

## 2.3 Process and Strategy

Effective, productive stakeholder engagement requires an understanding of who your stakeholders are, why they are important, and what issues are important to you both. When developing its external engagement process, AEP identified its stakeholders with help from Ceres and others. In some cases, the company used survey tools to identify concerns that would later inform content of stakeholder meetings. In others, the dialogue was a natural extension of ongoing discussions.

As relationships and trust grew stronger, it was not uncommon for an NGO to directly call the CEO or another senior executive to talk about an issue. AEP would also reach out to stakeholders to provide a "heads up" on an announcement or action the company was taking that would be of interest to certain groups. For example, when AEP announced plans to build new ultra-supercritical coal-fueled power plants, the company arranged a conference call with a small group of NGOs to share the announcement in advance. In another case, the discussion about climate change led to an invitation for NGOs to visit AEP's Mountaineer Plant in West Virginia to see the world's first validation project of carbon capture and storage on an existing coal plant.

---

[5]Personal communication, AEP stakeholder meeting, November 13, 2013.

Stakeholders can influence public policy and regulatory outcomes that directly affect the company. This happens at the national level and in local jurisdictions where NGOs have increasing influence and financial backing to support their causes.

AEP's philosophy has always been to work with its stakeholders in good faith to try to find common ground. The company believes that being honest and candid about the decisions it makes and the reasons for them is a prerequisite for successful engagement. AEP's experience shows that stakeholders will respond in kind—even when views differ sharply.

In 2006, energy efficiency programs were in place in two AEP jurisdictions—Texas and Kentucky—where they were mandated. Environmental groups pushed AEP to expand those programs to other states. They believed that energy efficiency was important to helping address climate change, a significant sustainability issue for AEP. At the time, coal-fueled electricity was relatively inexpensive and few thought there would be regulatory support for new programs that would raise customer rates. At the same time, the cost of new environmental controls on coal plants was beginning to hit customer bills. There was little support for adding to that burden.

Following more than a year of dialogue with stakeholders, then CEO Mike Morris set a goal to reduce electricity demand by 1,000 megawatts (MW) by the end of 2012. A year later, he added a goal to reduce energy consumption by 2,500,000 megawatt-hours (MWh) during the same time frame. Energy efficiency was the pathway to achieve those goals. AEP's strategy was to work closely with coalitions of stakeholders who were, in many cases, interveners in state-level regulatory cases. This collaboration was particularly important as states began establishing energy efficiency and renewable energy mandates. Working together ensured a good outcome for AEP and its stakeholders and opened the door to a new business opportunity for the company. AEP exceeded both the energy efficiency and energy demand goals.

Having a clear understanding of the issues that are most important to AEP and its stakeholders is essential to building credible, trusting relationships. Since 2007, AEP has reached out to stakeholders each year to ensure their priorities have not changed and to learn if something new has emerged. The company uses survey tools, conference calls, and other means to do this.

At AEP, the top issues to stakeholders have consistently been coal, energy efficiency, utility business model, and climate policy. More recently, water risks, political and lobbying activities, and new technologies (i.e., distributed generation and micro-grids) have infiltrated the discussions. Often, issues the company is questioned about are triggered by recent events. For example, following the 2010 *Citizens United* Supreme Court decision allowing companies to make political contributions to candidates, advocates began asking for greater disclosure on that issue. In response, AEP began to publicly disclose its political contributions and lobbying expenses, as well as the portion of trade group dues that were used for lobbying activities.

In 2012, AEP conducted its first formal materiality assessment to understand the sustainability issues most critical to the company and its stakeholders.

Materiality is most commonly associated with financial reporting. It is an accounting standard that provides a threshold for disclosure in financial reports. Knowing what issues are most important to the company and its stakeholders is also a key factor when disclosing sustainability performance. AEP considers material issues to be those that have affected or that are reasonably likely to affect the company's reputation, liquidity, capital resources, or results of operations. Material issues can also include those that stakeholders consider important to their interests and to AEP's sustainability. Conducting a materiality assessment allowed for consideration of what was important from the perspective of both the stakeholder and the company, creating opportunities for external and internal engagement.

The company used a consultant to help guide the process and develop a survey instrument that would identify the top 15 material issues. At the time, an analysis of AEP's sustainability Web site (www.AEPsustainability.com) revealed that the company was reporting on more than 80 issues or aspects of issues. Internal resources could not sustain that level of reporting nor was it necessary. The company used the GRI, International Integrated Reporting Council (IIRC), and the Sustainability Accounting Standards Board (SASB) definitions of materiality to guide the assessment. SASB and IIRC were still in the drafting phases but were far enough along to understand their unique perspectives. Considering all three gave AEP a broader view of what external stakeholders might be seeking in public disclosure. In the end, AEP defined materiality as a combination of financial and sustainability thresholds rather than choosing one over another.

Gaining a better understanding of what was relevant to investors was an important objective of the project. To that end, AEP engaged an independent sustainability investment analytics firm. AEP has produced an integrated report of financial and non-financial performance since 2010, making this perspective invaluable to the report process and focus.

Through a series of internal workshops with subject matter experts, the company narrowed the list from 80 to 36 issues or aspects of issues and developed a survey tool. Throughout the process, the team tested issues and their definitions with internal experts and with the financial analyst, and there were sometimes disagreement on an issue's importance. While not necessarily eliminating the issue, it prompted the team to rethink why it was important and to whom.

The team developed a targeted list of internal and external stakeholders that included AEP board members, management, subject matter experts, NGOs, suppliers, customers, regulators, investors, trade organizations, and labor. The survey was sent to 250 stakeholders with a two-week response deadline. AEP received a 54 % response rate.

The analysis provided clarity around the key material issues but missed the mark on some. In developing a materiality assessment, companies should exercise judgment when deciding what makes the list and what does not. For example, the survey indicated that political contributions and lobbying activity was a low priority issue for most stakeholders. However, AEP received shareholder requests regarding this issue several times. In this case, it was a material issue for the company which warranted continued reporting in the annual Corporate Accountability Report.

To validate the results of the assessment, the team reviewed the list of issues to determine if they aligned with the company's risks, as managed by the Enterprise Risk Management group. This provided assurance that the results were accurate and aligned. The final list of material issues was 17—a much more manageable and relevant level of disclosure (Figs. 2.2 and 2.3).

The materiality assessment took six months to complete. It is not necessary to do this every year, but AEP seeks to test the relevance of the issues and identify any new issues that might emerge annually. In late 2013, AEP assembled subject matter experts for a half-day workshop for a review. Interestingly, three issues not previously ranked as priority (but were in the original survey) rose to the top. These were customer relationships/satisfaction, engaging employees and market competitiveness. These priority issues reflected the strategies put in place in 2013 to support AEP's strategy for growth. After external stakeholders validated the

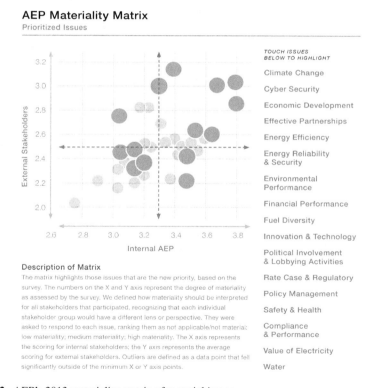

**Fig. 2.2** AEP's 2013 materiality matrix of material issues

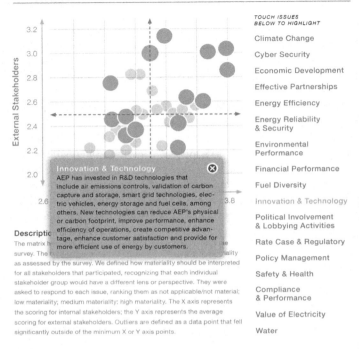

**Fig. 2.3** AEP's materiality matrix allowed online users to learn how each material issue was defined

importance of these issues, AEP added them to its 2014 Corporate Accountability Report.

## 2.4 Applying the Stakeholder Engagement Model Internally

The employee culture survey AEP conducted in 2012 was the equivalent of a materiality assessment of the company itself. While the materiality assessment included input from some employees, it was limited to technical experts and senior managers. AEP needed to conduct a broader survey with employees to understand their issues and perspectives. The survey was intended to measure the health of the organization and identify areas of strength and opportunities for improvement. The mixed results immediately prompted senior management to further engage employees to better understand the strengths and weaknesses of AEP's culture.

The survey found that AEP has a strong culture of safety and compliance with rules and standards and a solid basis of committed and experienced employees who care about the company and customers. The survey also identified several disconnects:

- Employees did not broadly understand the direction and strategy of the company, missing opportunities to effectively turn that into specific goals and targets.
- Employees did not feel sufficiently involved and empowered to take on tough challenges.
- There was room to improve how the company "keeps score" and provides recognition and rewards for top performers and holds weak performers accountable.
- Employees wanted more bottom-up innovation and more knowledge-sharing and cross-functional collaboration.

Management acted quickly. From the survey results, AEP identified four areas of focus to improve the health of the company's culture:

- Strategic alignment,
- Leadership development,
- Performance recognition and accountability, and
- Employee engagement.

Senior executives sought to conduct employee focus groups to learn more and gather ideas for improvement. The sustainability team was asked to apply the stakeholder engagement model internally. At the same time, a core culture team was formed (that included sustainability) to begin developing a culture road map. A full-time position was created to steward the initiatives and ensure ongoing progress, reflecting the importance of employee culture to AEP.

In 2013, 90 employee focus group meetings were held. Participants represented every level (except executive management), job type, demographic make-up, and cultural diversity. Employees were given an opportunity to volunteer to participate and regional human resources managers and business unit leaders reviewed the teams to ensure balance and diversity. Throughout the course of the year, more than 1,000 employees took part in online or in-person focus groups, nearly 6 % of AEP's full-time employee population.

Additional employees were trained as facilitators and given a guide for the two-hour meetings. Having employees lead these sessions created new openings for leaders to provide a development opportunity for some of their top performers. The discussions concentrated on the four focus areas of AEP's culture initiative. The facilitators worked in pairs and were assigned outside of their normal work area, adding a level of anonymity intended to foster an environment of safety and openness. Stakeholder meetings do not necessarily end with participants working together on projects every day. Stakeholders also are not concerned about the potential for manager retaliation or job security. While many of the external engagement principles applied directly to the employee focus group process, certain additional considerations were needed. Given that employees would continue to have to work together, and may be concerned about the response to their comments in the focus groups, AEP implemented a few modifications.

Participants were only asked to share their first names and years of service to protect their anonymity. Not even the facilitators were told who would be in their sessions. The comments were captured by one of the facilitators, and the result was a treasure trove of rich data that helped AEP prioritize initiatives and inform the employees' role in the company's business strategy. One of the reasons for providing a "safe" environment for candor about AEP's culture was that the company wanted honest feedback (Fig. 2.4).

When the meetings were complete, the team compiled the information gathered and used it to develop a culture road map, including a culture goal tied to incentive compensation. In addition to concerns and complaints, employees shared hundreds of ideas for improvement. For example, a majority of employees said that a simple "thank you" from their supervisor for a job well done was more meaningful and motivating than a gift card. That was in contrast to an employee who received little recognition for his 35-year service anniversary other than a certificate of appreciation from the CEO; his immediate supervisor did not take the time to say thank you for all those years of service. These starkly contrasting examples of what worked well and what the company knew it had to change provided insight that continues to guide AEP's culture journey.

Using the stakeholder engagement model to engage employees gave AEP the framework it needed to launch focus groups more easily and swiftly. It sent a

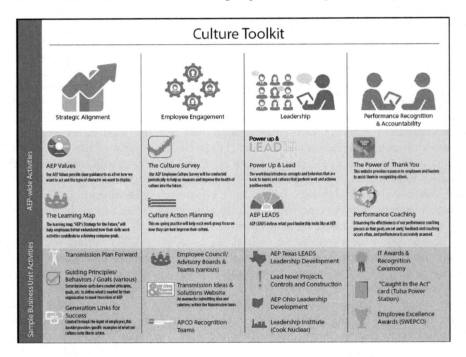

**Fig. 2.4** AEP's culture toolkit gives leaders and employees tools to support the company's four focus areas—strategic alignment, employee engagement, leadership, and performance recognition and appreciation

**Fig. 2.5** AEP's Engage to Gain program encouraged employee engagement to identify cost savings and process improvements. It included a potential financial reward for all employees

signal to employees that management was listening and wanted their ideas. This was particularly vital to a continuous improvement initiative also under way. Employee teams were formed to identify cost reduction opportunities as well as process and efficiency improvements. The "lean" process places a high emphasis on engagement and culture which complemented AEP's broader culture initiatives. The employee-led continuous improvement efforts further underscored the company's commitment to internal engagement.

In 2013, AEP established a gain-sharing program called Engage to Gain. Facing some significant business challenges, the company needed an engaged work force to ensure long-term financial and operational success, obviously critical aspects of being sustainable. AEP committed to employees that savings accrued beyond $200 million that year would be split 50-50 with employees (excluding senior management). Employees had to make the business case for the cuts, cost reductions, or process improvements and show that they could be sustained. At the end of 2013, employees identified more than $200 million and, as a result, every employee received a check for the maximum payout of $1000 (Fig. 2.5).

The strategy to involve employees was enhanced by the stakeholder engagement process that was already in place. With few modifications, the company directly engaged more than 1000 employees through focus groups and many more through the Engage to Gain program.

## 2.5 Tips for Successful Engagement

### 2.5.1 External Engagement

Successful stakeholder engagement requires a strategy and a plan. Once AEP decided that stakeholder engagement would be integral to its performance reporting and provide an avenue for more open dialogue, the company began to map its stakeholders and issues. It also requires collaboration across the organization for both internal and external engagement. This helps break down silos, creates greater understanding of business issues and opportunities, and fosters an environment of collaboration.

The first step is to identify the most important issues to the company and its stakeholders. It is also important to determine how the engagement process can have the greatest impact on business strategies and operations. Through internal discussions and working with Ceres, AEP identified the issues that would launch the first stakeholder meeting. Later, new issues would lead to follow-up engagement. With a robust process in place, the issues became readily apparent and began to build upon one another.

Next, it is important to know which stakeholders can have the most impact on your company's operations. For example, does the stakeholder have influence that can affect a regulatory proceeding or public policy outcome? If so, that is someone you would want at the table. Once that is complete, you can map your stakeholders to the issues.

AEP nearly always contracted with a third-party facilitator for its stakeholder meetings. Having a neutral, third party who understands your business as well as the principles of sustainability allows for more candid and constructive dialogue. It is important for the facilitator to get to know the company's senior team, understand the issues, and know who the stakeholders are. This creates trust and a comfort level that fosters an environment that allows for candid discussion. The facilitator ensures the conversation stays on track and that everyone is heard.

The reason a third-party facilitator is beneficial is that it allows AEP to be part of the discussion rather than leading the conversation. It leads to more meaningful dialogue and eliminates the awkwardness of cutting a conversation short or having to cut off a key stakeholder to stay on track. It helps to preserve the relationship AEP has with its stakeholders. Only occasionally would the sustainability team facilitate a meeting. These instances usually involved a conference call that required immediate response or discussion of an important issue or the facilitation of small group meetings on specific localized issues with subject matter experts. This approach worked smoothly for those circumstances.

Following every stakeholder meeting, it is important to follow-up with your stakeholders. They want to know what actions the company is taking or will consider taking. AEP and Ceres (and later SustainAbility, another consultant, which served as a facilitator for many years) took meeting notes that were combined and shared with the stakeholder team. The notes did not identify who said what, only who was at the table and the issues that were discussed.

In addition to sharing meeting notes, AEP would inform the stakeholders if the meeting led to a new goal or sustainability commitment. The company kept track of every request stakeholders made and the disposition of those requests. AEP communicated those results to the stakeholders, including which requests would or would not be acted upon and why. It was a transparent exchange that helped to build credibility and trust.

In 2010, AEP convened a meeting with its coal suppliers and the environmental community. Following a survey of the suppliers' environmental, safety, and health performance, AEP shared the results with both groups. It was also the first known survey of the coal industry with this focus. And, it helped AEP identify the percentage of coal in its supply chain that came from mountaintop removal mining practices—an issue of much stakeholder discussion. The survey was an important tool for AEP to understand the environmental and social performance of its coal suppliers, since the company was one of the largest consumers of coal. For the suppliers, it validated their performance claims, while it gave NGOs new insight into the industry's practices.

AEP understood that bringing NGOs together with coal suppliers would not be easy, but the company was in a unique position to convene such a meeting. AEP had proven its commitment to engagement and transparency and had developed good relationships with many of its stakeholders. It was the first time the two

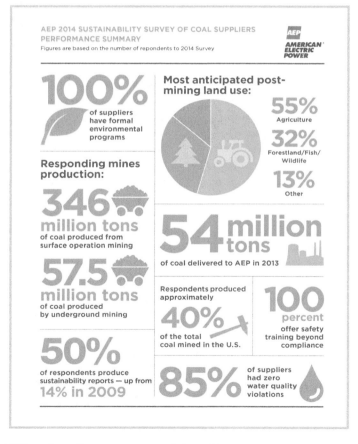

**Fig. 2.6** AEP's coal supplier survey summary infographic

groups came face-to-face for a dialogue on issues that, heretofore, they had only responded to in the media or through legal proceedings.

It was a rocky start. When the stakeholders first arrived for a welcome dinner at AEP, there was little to no mingling among them. The next morning they came into the meeting together, laughing and talking to each other. When asked what changed, they said they spent time at the hotel bar the previous evening talking to each other. Although those relationships remain contentious, there emerged a new dimension to them that never would have otherwise occurred (Fig. 2.6).

Failure to engage can potentially damage a company's reputation, create financial risk, threaten public trust, and put a company's license to operate at risk. Today, shareholders are not only taking their concerns or demands to the company, but also to other influential shareholders to increase pressure for change. The sphere of influence that NGOs have today is broader, and their financial backing is stronger. Companies have more to lose than ever if they fail or refuse to engage.

**Tips for Successful External Engagement**

- Have an icebreaker that allows people to get to know each other.
- Use a third-party facilitator.
- Arrange seating so that stakeholders and company representatives are interspersed, avoiding an "us versus them" standoff around the table.
- Ensure everyone is heard; rely on the facilitator to draw everyone into the conversation.
- Provide access to the CEO for a one-hour open Q&A. Stakeholders covet access to senior management; providing it helps to build respect, trust, and credibility.
- Follow up to stay connected.

## 2.5.2 Internal Engagement

Collaboration and management commitment are critical elements of successful employee engagement. Where employees are empowered to offer ideas, lead teams, and find solutions, the outcome is invariably a win-win for the organization. This approach can also inspire innovation, break down silos, and lead to new business efficiencies, process improvements, waste reduction, cost savings, and higher employee morale.

The key to doing it right is to collaborate with those who are inside business units, who have their finger on the pulse of different parts of the company—the company's social influencers who are not always managers but know what is going on. For example, when starting the process to organize employee focus groups, a partnership between AEP's sustainability team and human resources (HR) was fundamental to ensuring a positive outcome. The HR team was better positioned to help identify employees who would be active and constructive contributors, however vocal they might be. They could also screen for performance issues that might preclude someone from participating. The sustainability team was critical for applying the methods from the external engagement process.

Corporate communications was another important partner as they were instrumental in developing and disseminating communications to support recruitment and reporting results. In most cases, AEP held "open enrollment" to allow all employees an opportunity to volunteer for the focus groups. Providing a degree of anonymity was important to foster an environment of candor during the meetings. By limiting access to the master list of participants, the employees could be confident that their comments would not be reported back to their supervisor. This was vital since the idea was to solicit honest, candid feedback about moving AEP's culture forward. A special e-mailbox was created to collect nominations to which only the sustainability team had access.

Getting the right mix of employees—diversity, types of jobs, levels in the organization, represented, exempt, etc.—is critical to be credible. For example, if the focus groups consisted of only managers and supervisors, it would have been a lopsided conversation with only one point of view. Here is where the partnership with the human resources and the business unit leaders was critical. They were able to help narrow the lists to workable numbers for each group. In the end, each group comprised a maximum of 15 employees and two trained employee facilitators.

In addition to the smaller groups, AEP formed a couple of "mega focus groups." The company brought together about 100 employees at once and split them into ten groups of ten. This was done twice—once during the initial round of meetings in early 2013 and again in the fall to focus on updating the company's values and purpose statement. AEP used this approach as a way to concentrate a larger number of meetings in a shorter period of time and to eliminate the need for more meetings.

During the mega focus group on values and purpose, the employees were told that senior executives would visit during the afternoon session, when the teams did their report outs on their group work. In this case, executives had started the work on the values and purpose statement but wanted to get input from employees.

Nearly the entire senior team—including the CEO—attended, sat with employees around the room, and later stayed to mingle and talk one-on-one with the employees. In follow-up communications with the employee team, many expressed appreciation for that opportunity.

The data collected from employees during the meetings were categorized by focus area (strategic alignment, employee engagement, leadership, and performance recognition and accountability). Reports for each focus area were compiled and shared with senior management. These reports informed development of the next year's culture road map, led to employee-led teams focused on engagement and employee recognition, helped to spawn the creation of culture committees across AEP, and prompted an overhaul of the performance and compensation management systems. Regular communications keep employees informed of the company's progress on all fronts.

Stakeholder engagement, like employee engagement, is a continuous process. Companies that veer off course from this commitment risk losing the trust, the credibility, and the "bank of good will" that engagement builds.

**Tips for Successful Internal Engagement**

- Use the external engagement model as the basis for engaging employees.
- Form partnerships with business unit leaders and the HR and corporate communications teams—they are valuable allies and advocates.
- Be clear about the desired outcomes. Create a facilitator's guide that drives the conversation in that direction and keeps it focused.
- Capture comments and ideas (anonymously). Gather the positive as well as the negative. This will be useful information to inform follow-up actions.
- Do not allow supervisors and their direct reports to be in the same session or candor will be inhibited.
- As eager as senior executives are to participate, do not let them. Employees will not be honest or forthcoming about their concerns with an executive present. If there is an occasion to bring in executives that is appropriate, warn employees ahead of time.
- Solicit participants from all levels of the organization, diversity of job types, and cultural diversity.
- Follow up. Employees want to know what will happen with the data collected from the focus groups. Communicate intended actions. When something occurs that was a result of employee input, recognize it. If you don't, employees will disengage.

## 2.6 How Sustainability and Culture Are Transforming AEP

The process and principles for conducting stakeholder meetings are equally effective in organizing employee focus groups. By talking with and listening to external stakeholders, AEP is able to consider different inputs to the decision-making process. Whether it comes from a customer, an employee, or an environmental group, companies need information to execute their business strategy effectively. Engaging employees is equally as valuable and often leads to new ideas, business opportunities, innovation, process improvements, or cost reductions that otherwise might not be known.

Following a roundtable with executives and thought leaders from other industries, AEP CEO Nick Akins said he enjoys such exchanges because they helped him to learn about what others are thinking and to hear new ideas. He added that by broadening his scope of engagement, he increases the chances of learning something new that might benefit AEP. The same is true when working with employees.

During the course of conducting 90 employee focus group meetings in 2013, AEP gathered substantial information and dozens of ideas that have since sparked a series of initiatives to strengthen the culture. Some of the actions have included:

- Creation of a corporate Culture Advisory Board to provide bottom-up feedback to management on a range of issues and to help shepherd culture initiatives across the company. Many business units have formed local culture groups, too. The board organized a "Culture Stand Up" in 2014 to unveil an updated set of company values and a new purpose statement. Often, companies will hold a stand-down when a serious event has occurred to refocus and prevent a recurrence. In this case, AEP's employees "stood up" for AEP's values (Figs. 2.7 and 2.8).

**Fig. 2.7** AEP held a "Culture Stand Up" in 2014 to encourage engagement around revised values and a new purpose statement

**Fig. 2.8** AEP's revised values and new purpose statement were developed with employees' input

- Employees developed an engagement framework to help leaders, and employees have more meaningful interactions. The framework was developed by an 18-member employee-led team and provides common sense tips for improved interactions at all levels. It is not prescriptive and ready for immediate use.

  During focus groups, employees said they wished their leaders better understood the impact their decisions had in the field. They challenged leaders to "walk a mile" in their shoes. Employees wanted to understand why certain decisions were made and that management understood the operational implications. In response, the framework was created to improve communications and interactions. The team further raised the bar by setting forth the expectation that employees will take an active role in engaging with each other, recognizing they have an equal responsibility for positive, effective engagement.
- Creation of a new leadership framework, called AEP LEADS, was designed to clearly define what makes a great leader at AEP. The engagement framework is one of the tools that support this effort (Figs. 2.9 and 2.10).

How does this tie to AEP's sustainability? During a meeting with external stakeholders in late 2013, AEP's CEO and other executives talked about the company's

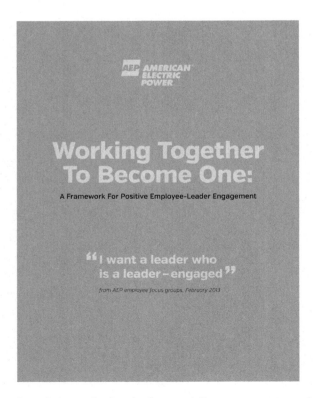

**Fig. 2.9** An employee-led team developed a framework for engagement targeted at all levels of the organization

**Fig. 2.10** AEP's new leadership framework aligns with the company's values and supports its culture journey

focus on culture to address the many different challenges it is facing. Nick Akins talked about his own personal commitment to develop a work force that is more collaborative, engaged, and entrepreneurial. He spoke about it in the context of the significant transformation AEP, and the electric utility industry is undergoing. "AEP cannot achieve its goal to be a sustainable company without its employees. Just as our stakeholder engagement is critical, a healthy internal culture is a business imperative," he said.

It was an investor who immediately understood the connection. He described the focus on culture as a measure of how well a company is managed. He added that the "tone at the top" is vital to this. It was the first time AEP and its stakeholders heard the connection between internal culture and corporate sustainability verbalized.

AEP frequently shares this account with employees to help them understand why culture is not solely an internal journey. It helps employees understand that external stakeholders, communities, and ecosystems have a vested interest in it as well.

## 2.7 Future Trends

According to a 2014 Aon Hewitt research report on global employee engagement, companies who engage their employees during challenging times "are more likely to emerge successful when the crisis is over. … These factors include focusing on long-term strategies, demanding measurable action, involving all stakeholders, understanding key employee segments, and broadening the range of assessment tools and analytics."[6] AEP's four focus areas on its culture road map align with these factors.

Employee engagement can have a direct impact on financial performance. A company that considers employees to be critical partners for future success recognizes the power of an engaged work force. Employees who are engaged can counteract threats to the company's brand, reputation, and image, for example. Conversely, when employees are not engaged they can damage reputation and profitability through passive sabotage. A healthy culture and an engaged work force are critical to AEP's future. Consider the success of AEP's 2013 gain-sharing program as an example of the power of engaged employees.

Many employees are also shareholders of the company and have a vested interest in AEP's success. It is one reason they are encouraged to be active contributors. In response, employees are innovating, streamlining work processes, eliminating waste, and working collaboratively. For example, as the Economic and Business Development team sought to leverage opportunities from shale gas growth in AEP's service territory, the company had to find new ways to expedite large-scale electric service to remote locations.

The challenge was whether AEP could deliver enough electricity to power the construction and startup of an expansive (and expensive) industrial plant that was going to process natural gas that drillers were capturing from the Utica Shale formation a mile beneath the surface in rural eastern Ohio. A lot was at stake: The facility represented increased sales for AEP, hundreds of jobs for local residents, and new tax revenues for schools and local governments.

Without a solution from AEP, the company would have been forced to use noisy and noxious diesel-powered generators requiring environmental permits, on-site fuel tanks, 24/7 operators, and tanker trucks, adding to local traffic congestion. Transmission engineers had been working on developing a new mobile skid-mounted substation for fast delivery, but it was still untested. With agreement from the customer to test it at the site, the mobile substation proved to be the solution the customer needed. In addition, it aligned with AEP's business objectives to provide speedy customer service and to embrace a business culture transformed for success in a competitive world.

### 2.7.1 What the Future Holds

- Ultra-transparency will give companies a competitive advantage.[7] Transparency is a basic expectation, and basic is no longer good enough.
- Proactive, real-time engagement with stakeholders will be critical to leverage opportunities and mitigate risks. Companies that do not engage will face

---

[6]Hewitt [1].

[7]May 2014 presentation to NAEM by Michael Muyot, President, CRD Analytics.

increased shareholder and consumer activism and poor employee morale. Stakeholders want to understand what companies are doing, as they do it, not after the fact. Employees want the same information.
- Intrapreneurs—those who think and act like owners of the company—will underpin a company's health. Turning entrepreneurship loose inside a company will help build buy-in and ownership.
- Being engaged means staying engaged. Internal and external stakeholders want an ongoing connection, not one-time outreach.
- Every interaction creates a new dimension to relationships. When people come together, their perceptions change and they often have a deeper understanding of each other's perspective. Internally or externally, they better understand the what, why, and how of business decisions and actions. Agree or disagree, the relationship is different and stronger.

There is no silver bullet when it comes to internal or external engagement. Companies may shy away from engaging externally for reasons that include unwanted public scrutiny, unpredictable outcomes, and fear of reprisal. However, a well-managed process for engagement can reduce these risks while unlocking the potential for greater collaboration, sharing of ideas and innovation. It can also help to improve a company's "social capital" by demonstrating a good-faith effort to build trusting, credible relationships. These attributes of stakeholder engagement, while hard to quantify, can be as valuable to a company as quarterly profits.

Internally, engaged employees may be hard to find, but it is a quest for discovery that no company can afford to ignore. During a time of major transformation within the electric utility industry, an engaged work force is vital to sustainable, profitable business. It is well-known that engaged employees will go the extra mile, are passionate about their work, are committed to serving customers, and are the ones who feel connected to the company. Culture change takes time, commitment, and deliberate action. But the return on investment is limitless when employees are empowered to contribute in meaningful ways, and leads to a more sustainable organization.

For AEP, internal and external engagement has become a natural course of how the company conducts its business.

## Reference

1. Hewitt A (2014) Trends in global employee engagement. Retrieved from: http://www.aon.com/attachments/thought-leadership/Trends_Global_Employee_Engagement_Final.pdf

# Chapter 3
# BC Hydro: How Demand-Side Management Can Help Utilities Achieve Triple Bottom Line Outcomes

**Brenda Goehring and Kristin Hanlon**

**Abstract** The conservation and efficient use of electricity at customer locations is the focus of demand-side management (DSM). The business case for DSM seems straightforward: DSM programs increase customer satisfaction and provide opportunities to lower electricity bills, reduce utility revenue requirements through offsetting more expensive supply-side generation options, and avoid the environmental impact of new generation. However, as the electric utility market and industry evolve over time, the business drivers for investing in DSM change in importance, resulting in the need to evolve the business case. BC Hydro's demand-side management program, Power Smart, celebrated its 25th anniversary in 2014 and has achieved 4100 GWh in energy savings over the period fiscal year 2008 to fiscal year 2015; which is enough to power almost 400,000 homes. Using case studies, interviews, and program performance, the authors track the successes and challenges with delivering Power Smart over the last 25 years and offer lessons for peer utilities to consider how DSM can complement utility sustainability initiatives based on BC Hydro's experience.

## 3.1 Introduction

Demand-side management (DSM) broadly defined is action taken by a utility to support its customers to conserve or use energy more efficiently. Some jurisdictions also consider actions to reduce the energy demand that the utility serves to be DSM, such as

---

B. Goehring (✉)
Corporate Policy and Reporting, BC Hydro, Vancouver, Canada
e-mail: Brenda.Goehring@bchydro.com

K. Hanlon
Strategic Planning, Conservation and Energy Management, BC Hydro, Vancouver, Canada

© Springer International Publishing Switzerland 2016
J. Fox (ed.), *Sustainable Electricity*, DOI 10.1007/978-3-319-28953-3_3

displacing all or part of a customer's load through distributed generation. From a sustainability perspective, DSM can bring all the key triple bottom line elements together. It has little to no environmental impact and can avoid the environmental impact of new generation. It is typically cost-effective compared to supply-side resources (electricity generation), and it has a number of social benefits such as electricity bill reduction. However, DSM is not always an obvious investment from a utility perspective. It can reduce utility revenue requirements, which can be challenging to rationalize to shareholders and it may require a long-term view from initial investment to realized electricity savings. It may also have a rate impact. DSM as a resource option has different costs, benefits, and risks compared to supply-side resource options, which take time to understand, analyze, and adopt. Finally, DSM involves working with customers to change their purchasing decision and behaviors to reduce electricity consumption. Doing this effectively involves innovative strategies and engagement skills that may be outside the experience traditionally found in utilities.

From BC Hydro's experience, there is opportunity for utilities to consider the role of DSM as a long-term resource to meet customer needs, avoid, or defer new generation or transmission and distribution additions with their associated capital cost and environmental impacts, while proactively engaging customers to participate and change behaviors. Through 25 years experiencing highs and lows in advancing its program, BC Hydro offers insights on how utilities can explore how to approach similar offerings to its customers. There is recognition that investor-owned utilities

**Fig. 3.1** Value proposition for DSM

may be challenged with rationalizing investments in DSM and that regulators may not have the tools to evaluate these offerings consistently. However, there are sound business drivers, policy, and regulatory considerations, as well as market share that can augment traditional customer offerings in a shifting landscape for utility providers. This chapter will seek to lay out the opportunity, challenges, and learnings that can inform this strategic business opportunity (Fig. 3.1).

## 3.2 The Context

### 3.2.1 Demand-Side Management at BC Hydro

BC Hydro's DSM plan consists of three main tools that are used to achieve energy savings: programs, codes and standards, and conservation rate structures. In addition to these three tools, there are a number of supporting initiatives such as public awareness and education, community engagement and advanced DSM strategies that provide a critical foundation to support the success of DSM.

- **Programs** are designed to address barriers to DSM, such as increasing the awareness of energy conservation and identifying opportunities to reduce consumption. The programs also improve the availability of DSM products and services and provide incentives to motivate customers to invest in conservation and efficiency if affordability is a barrier. BC Hydro has a wide-ranging approach to DSM, involving retailers, distributors, service providers, and manufacturers to maximize Power Smart's reach, manage costs, and support the adoption of energy-efficient products in BC. BC Hydro has approximately 10 broad programs in market, many with multiple offers, targeting the residential, commercial and industrial customer classes. Programs are designed to support codes and standards and conservation rates, which are described further below.
- **Codes and Standards** are public policy instruments enacted by governments to influence DSM. Examples include building codes, DSM regulations, tax measures and local government zoning and building permitting processes. BC Hydro works with federal, provincial, and local governments to support new codes and standard development across multiple customer sectors as well as to support enforcement and compliance efforts.
- **Conservation Rate Structures** are inclining block (stepped) rate structures. Conservation rate structures were in place for BC Hydro's residential, commercial, and industrial customers over the period fiscal year 2009 to fiscal year 2015.

As shown in Fig. 3.2, BC Hydro has achieved 85% of future energy savings, from programs and codes and standards, with the remainder coming from conservation rates. The total amount of savings is spread relatively evenly among the residential, commercial, and industrial customer classes.

BC Hydro's investment in DSM is guided by provincial policy and legislation and determined through BC Hydro's integrated resource planning (IRP) process which develops a long-term action plan to meet customers' electricity needs over 20 years.

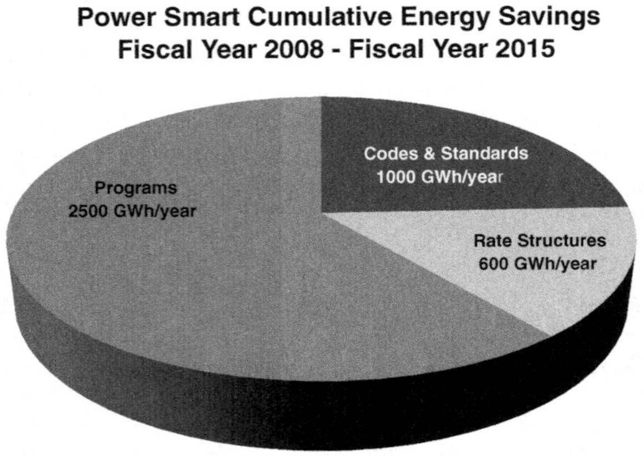

**Power Smart Cumulative Energy Savings
Fiscal Year 2008 - Fiscal Year 2015**

- Programs: 2500 GWh/year
- Codes & Standards: 1000 GWh/year
- Rate Structures: 600 GWh/year

**Fig. 3.2** BC Hydro's Power Smart energy savings—fiscal year 2008 to fiscal year 2015

The BC *Clean Energy Act*[1] enacted by the BC Government in 2010 sets out 16 British Columbia energy objectives, of which one is for BC Hydro to meet 66 % of all new power demand through conservation by 2020. The IRP is required to respond to these energy objectives as well as consider the cost-effective resource actions to meet customer reliability requirements and address environmental concerns.

BC Hydro's DSM initiatives have been known as Power Smart, which is a brand developed and trademarked by BC Hydro. Power Smart has been in existence for over 25 years coinciding with the launch of DSM and today has as much brand recognition among customers as the company itself–to the extent that the brand was adopted in 2015 as BC Hydro's company brand with the tagline *Smart About Power in All We Do*.

## 3.3 Power Smart's 25-Year History

BC Hydro started its Power Smart program in 1989 and celebrated its 25th anniversary in 2014. Since fiscal year 2008, it has achieved approximately 4100 GWh in energy savings from programs, codes and standards, and conservation rate structures. While it is an embedded part of BC Hydro's business today, Power Smart has had its share of ups and downs over its 25-year history (Fig. 3.3).

### 3.3.1 1980s and Early 1990s: Power Smart Launch

Power Smart was conceived at BC Hydro during the late 1970s and the early 1980s when international oil supplies were constrained and a recession was underway, prompting views that conservation of natural resources was necessary to

---

[1]https://www.leg.bc.ca/39th2nd/1st_read/gov17-1.htm.

3 BC Hydro: How Demand-Side Management Can Help Utilities … 35

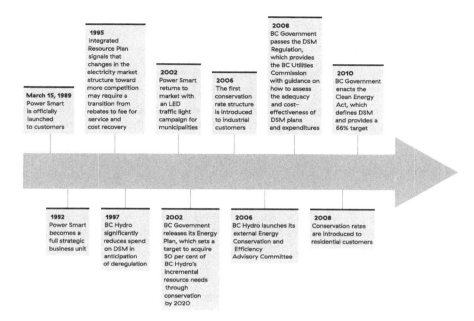

Fig. 3.3 Power Smart's 25-year history

create economic stability and to manage inflation. These influences also extended beyond oil to electricity. In parallel, there was a growing environmental movement among the public, and citizens expressed a desire to be consulted on public policy decisions and for the government to ensure that BC Hydro took non-economic factors into consideration when developing new projects.

The Province of British Columbia in reaction to these influences shifted away from an asset-build perspective and introduced a new 1980 Energy Plan, which emphasized energy security. Through this policy, BC Hydro introduced its first industrial and commercial DSM programs, which represented a shift from previous energy policies focused on building out provincial generation and transmission facilities.[2]

Before Power Smart was officially launched, it was a modest initiative where technical staff researched and promoted energy-efficient lighting, home retrofits, and high-efficiency motors to customers. The vision for launching Power Smart came from Larry Bell, president and CEO of BC Hydro from 1987 to 1991. Bell was an economist who believed in the cost-effectiveness of DSM, particularly as a means to offset potential rate increases for customers, as well as providing price signals to encourage the efficient use of electricity. Bell is widely credited internally and externally as championing the start of Power Smart and reinvigorating it in later years. Bell thought that conservation was simply the right thing to do.

---

[2]B.C. Ministry of Energy and Mines Discussion Paper, *The Evolution of British Columbia's Industrial Electricity Policy*, p. 4.

> "You've got something driven by economics. It rewards the customer, it rewards the utility, and it reinforces a common ethic about not wasting things. The stars are aligned when that happens. It seemed to me it was just the right thing to do. It made sense economically and ethically. I guess it became part of the green ethic that pervades society today, but in those early days it was mostly about waste, not throwing things away. Our society has never endorsed that."[3]

While Bell was a supporter of DSM, there were two obstacles that he had to overcome to expand the effort within BC Hydro. First, there were few people within the organization that were working on DSM and they were largely technical staff. Second, it became apparent that the barriers to customers investing in DSM were not just economic.

> "You had a superb engineering organization with highly motivated people whose orientation was to build, so shifting gears was difficult. And the increment in savings was not that important for people's [customer's] budget. With the industrials, electricity was very inexpensive and, unfortunately, we had built an industry that had made perfectly rational decisions about how to build their factories based on the fact that energy was cheap."[4]

One answer to both of these issues was to hire new employees with marketing expertise to promote energy efficiency to BC Hydro customers. Employees with marketing expertise were able to translate technical concepts and issues in a way that customers would understand and contribute significantly to program design so that the Power Smart message would reach its desired audience. One of the first work tasks was to develop the brand for DSM at BC Hydro, which was when the Power Smart name was first conceived.

Power Smart was officially launched on March 15, 1989, which included a commitment at that time to a 10-year investment of 225 million dollars. Initial programs that were launched included the following: Refrigerator Buy-Back, Refrigerator Efficiency, High-Efficiency Motors, Water Heater Blanket, and Energy-Saving Fluorescent Lamps. (Fig. 3.4)

### 3.3.2 Late 1990s and Early 2000s: Utilities Move Away from DSM

During the 1990s, electricity policy in North America shifted to assessing options for market design, particularly looking at the deregulation of electricity markets as a means to open trade and mitigate rate increases. Customers wanted the ability to exercise supplier choice through open access transmission tariffs to secure inexpensive market supply that had been built in the past decade and, with a downturned economy, had created electricity market surpluses.

The mid- to late 1990s were extremely challenging for Power Smart as conservation, and DSM programs were effectively abandoned; program budgets and

---

[3]BC Hydro Power Pioneers with Kerry Gold, *Power Smart: How a Simple Idea Revolutionized the Way We Use Electricity*, Fig. 1 Publishing, Vancouver, 2014, p. 8.
[4]Ibid, p. 12.

**Fig. 3.4** Power Smart marketing campaigns over 25 years

resources were reassigned elsewhere within BC Hydro, particularly toward new products and services that could be recovered for a fee from customers. This was consistent with North American utilities in general who were preparing for deregulation of the industry. Annual Power Smart budgets dropped from tens of millions of dollars to a million dollars. Publicly the message was that Power Smart had been successful in seeding the market in DSM and BC Hydro no longer needed to be active in promoting DSM. DSM was a resource in BC Hydro's long-term resource plan, but no longer needed active investment.

### 3.3.3 Early 2000s to Present: Rebuilding and Sustaining Power Smart at BC Hydro

In the early 2000s, the Province of BC continued its intervention in market design and elected to not deregulate after observing similar actions in other jurisdictions and instead sought to freeze rates to support economic goals and industry retention. BC Hydro's primary strategy for meeting customer demand was to develop new gas plants on Vancouver Island. Investment in DSM continued to be a lower priority than supply-side resources.

However, in 2001, Larry Bell returned to BC Hydro for a second time as Chair and CEO and stayed in that role for six years. When Bell returned he asked to meet the head of Power Smart at the time. Power Smart still existed as a brand owned by BC Hydro, but little or no investment was being made to promote DSM programs. Bev van Ruyven was head of Key Accounts and was selected to meet Larry Bell because she had been trying to advance a business case for LED traffic lights that was desired by BC Hydro's municipal customers. Larry offered Bev a challenge–Prove there was still a DSM gap in the market that Power Smart could fill and if so, he was prepared to invest in DSM again. In addition to the LED traffic lights and Christmas LED lights, Bev led a team to develop a pilot program on Vancouver Island to trial a full suite of programs that had been in the market in the early 1990s, such as the Refrigerator Buy-Back program, Home Energy Upgrade program, and a Lighting program based on compact fluorescent light bulbs. The pilot program proved to be very successful and Bev was then asked to develop a long-term plan for DSM. Power Smart saw a renewal at BC Hydro between 2002 and 2003. Bev Van Ruyven, Senior Vice President, Customer Care & Conservation was recognized as a new champion for Power Smart. She recalls the experience of restarting Power Smart after a several year hiatus.

> "It was very challenging to bring Power Smart back, although we did it. One thing that helped immensely during this period was that we didn't lose the intellectual capital within the organization on DSM, although they were redeployed elsewhere within the organization. So we were able to rely on these skilled individuals again once we had a mandate to renew Power Smart. Another key factor during this period was that we had internal support from our President and CEO to invest the necessary resources in DSM. If there was less support for DSM during this period–to do it off the sides of our desks without a concerted effort–it would have been very challenging to make the inroads that we did." [5]

Even with internal support for Power Smart, it was not an easy process to relaunch programs in the market. Restarting and rebuilding relationships with retailers, contractors, and trade allies was particularly challenging. There was a lack of trust within the marketplace whether BC Hydro was committed to Power Smart over the long term. It took a number of years to recover from the exit from the marketplace in the 1990s. The Power Smart team relied on existing internal expertise, existing customer relationships and previously successful strategies for rebuilding the program.

Two other key developments during this period served to further embed DSM as part of BC Hydro's business. First, the Government of BC evolved its energy policy over this period, which provided support for sustainable electricity generation in general and DSM as a tool to meet sustainability and climate change goals. A 2002 Energy Plan set out objectives for BC Hydro with respect to conservation measures, notably to set conservation rates for all customer classes. The first provincial Climate Action Plan was published in early 2007 with new statutory and regulatory requirements, coupled with a carbon tax to reduce provincial GHG emissions.[6] Also in 2007, the provincial energy plan was updated and provided a requirement for BC Hydro to meet 50 % of its incremental resource needs through conservation by 2020.[7] By 2010, the BC *Clean Energy Act* was enacted, which further stretched this conservation target to 66 %.[8]

The second development over this period was the re-emergence of long-term resource planning at BC Hydro. From early on, integrated resource plans took the view that DSM was one of the available resource options to meet customer demand and accordingly was included with supply-side resources in portfolios of resource options. In BC Hydro's 2008 integrated resource plan, DSM was the first

---

[5]Personal Communication—June 20, 2014.

[6]http://www.gov.bc.ca/premier/attachments/climate_action_plan.pdf. 2007.

[7]http://www.energyplan.gov.bc.ca/PDF/BC_Energy_Plan.pdf. 2008.

[8]https://www.leg.bc.ca/39th2nd/1st_read/gov17-1.htm. 2010.

resource of choice to fill the load-resource gap. This strategy of conserving first became a cornerstone of subsequent resource plans and is also embedded within BC Hydro's Corporate Environmental Responsibility Policy.[9] In taking this approach, BC Hydro sought to defer or avoid new generation requirements given their large capital and operating costs, the associated environmental footprint, and the increasingly challenging regulatory and social landscape to permit and build new assets. It was prudent to explore DSM given the need for smaller scale supply increments following what is often called "lumpy" supply additions (when large increments of generation are added to the system all at one time) of hydroelectricity BC Hydro experienced when a new facility came online.

Over the last 25 years, it is clear that the results of investing in DSM reaped the rewards of sustainment, and even further, were a cornerstone of the government energy and environmental policies and regulatory framework. Examining how Power Smart was developed and evaluated, and its successes and challenges at a strategic and tactical level can support application in parallel utility plans.

## 3.4 Power Smart Program Impacts

In evaluating the impacts of the program, one key metric for measuring the impact of Power Smart at BC Hydro is reported cumulative energy savings. Since fiscal year 2008, BC Hydro has saved approximately 4100 GWh from Power Smart programs.

To put this into perspective, 4100 GWh of energy savings is roughly equivalent to the size of a new 750 MW natural gas generation facility, including the related transmission and distribution lines that would be required to deliver the electricity to the load. The investment in Power Smart also avoids the entire cost of the alternative investment and avoids the environmental impact to land and water of new generation, transmission, and distribution; and the challenging landscape for permitting of these assets.

Looking forward, BC Hydro will continue investments targeted to meet the 66% goal in fiscal year 2021 (the year of the BC government's DSM policy objective). Again, a key benefit of the investment is to defer investing in new generation resources. The net levelized Total Resource Cost[10] of BC Hydro's DSM plan is approximately $23/MWh, including energy savings from conservation rates and codes and standards as well as programs. Compared to generation resource alternatives analyzed in BC Hydro's 2013 Integrated Resource

---

[9]https://www.bchydro.com/about/sustainability/environmental_responsibility/environmental_policy.html.

[10]Total Resource Cost is a standard metric to evaluate DSM cost-effectiveness. It is a benefit–cost test, which looks at both utility and customer costs and benefits.

Plan, incremental requirements are estimated to cost $85–$100/MWh on the margin; therefore, there can be considerable savings to ratepayers with investing in DSM.[11]

In addition to these economic and environmental benefits, BC Hydro also realizes other important benefits through its Power Smart program:

- Customer satisfaction

Internal research conducted by BC Hydro over many years has indicated that the Power Smart program and brand contribute positively to overall customer satisfaction levels. As mentioned earlier, this relationship led to an opportunity to use Power Smart as BC Hydro's broader corporate brand.

DSM as a driver of customer satisfaction is supported by qualitative evidence from BC Hydro's Key Account Managers (KAMs), who are responsible for managing the relationships with BC Hydro's largest industrial and commercial customers. Customers seek out opportunities to make investments in DSM since they help to improve customers' bottom lines, make them more competitive in their markets, and provide synergies with a firm's reputation, marketing and branding efforts.

In addition, BC Hydro has found that through active program participation, customers also become more knowledgeable and seek to identify new energy-efficient and conservation opportunities to evaluate and pursue, often with positive outcomes that lead to new program research and development. Utility-customer partnerships become more common when both parties are seeking to manage the input costs of electricity into a firm's operations. This can be of strategic importance in a competitive marketplace.

- Increasing Stakeholder Expectations

One of the themes that emerged from BC Hydro's IRP consultation is that customers want to see DSM potential fully explored before embarking on new generation or transmission and distribution projects. In a jurisdiction where there are strong values around environmental protection, this is not surprising.

The conversation on whether BC Hydro has considered the appropriate amount of DSM in its planning for new projects also occurs within the regulatory process. There, DSM is explicitly considered to assess whether a new generation or transmission and distribution project is needed or could be substantially deferred from its proposed commercial operation date. It is anticipated that this will increasingly become a trend in North America by stakeholders who critique resource plans that illustrate gaps in making tradeoffs between a "build or conserve-first" approach.

- Economic development benefits

BC Hydro is guided by the BC *Clean Energy Act*, which requires BC Hydro to respond to 16 provincial energy objectives in its IRP process. One of the energy

---

[11] https://www.bchydro.com/content/dam/BCHydro/customer-portal/documents/corporate/regulatory-planning-documents/integrated-resource-plans/current-plan/0000-nov-2013-irp-summary.pdf.

objectives is to "encourage economic development and the creation and retention of jobs."[12] BC Hydro considered economic development benefits in its IRP process to respond to this objective.

BC Hydro designs its DSM program offerings through existing channels in the market to align with customer's current needs and values, which is one way to realize economic development benefits. For example, a key underpinning of program design is to leverage retailers, customer associations, and thought leaders to support key touch points to foster greater penetration and success. A program promoting new DSM lighting products can be supported through multiple avenues such as: a manufacturers' product offerings can be endorsed under the Power Smart brand; retailers can host educational point-of-sale seminars to augment customer knowledge and adoption; associations sponsor conference speakers and/or pilot installations to support program take-up, service contractors supplement technical knowledge and technology use; and program evaluations can be conducted to provide feedback on energy and bill savings to support repeat adoption. Enabling the DSM supply chain, upstream and downstream, has a positive effect on the secondary marketplace which services utility customers, allowing the utility to focus on their core service offerings.

As evidence of this, a recent study of the Canadian energy efficiency industry has found that energy efficiency significantly increases GDP and stimulates growth in employment. In particular, for every dollar invested in electricity DSM programs, there is a 4.8× increase in Canadian GDP. Similarly, investment in electricity DSM helps to spur a net increase in employment of over 650,000 job-years in Canada.[13]

## 3.5 Learnings

In stepping back and evaluating the investment in this program from a triple bottom line perspective, the following themes are evident.

### 3.5.1 Building a Base of Social Support

One of the factors that has been important in sustaining Power Smart has been engaging its customers and stakeholders at a number of different touch points. When looking at the role of engagement—with stakeholders, First Nations,

---

[12]https://www.bchydro.com/content/dam/BCHydro/customer-portal/documents/corporate/regulatory-planning-documents/integrated-resource-plans/current-plan/0000-nov-2013-irp-summary.pdf.

[13]Acadia Center, *Energy Efficiency: Engine of Economic Growth in Canada*, March 2014. http://acadiacenter.org/wp-content/uploads/2014/11/ENEAcadiaCenter_EnergyEfficiencyEngineofEconomicGrowthinCanada_EN_FINAL_2014_1114.pdf.

customers, and employees—it helps to see how these contributed toward to the program's design and outcomes. BC Hydro has a number of key strategies to engage with First Nations and its customers/stakeholders on DSM:

1. Use Key Account Managers (KAMs) to educate BC Hydro's largest customers on energy savings opportunities.

   BC Hydro has over 750 customers who are considered key accounts, and who work with KAMs. While the strategy for the KAMs has evolved over the years, it has a foundation to deliver on conservation targets through the range of Power Smart Programs and the delivery of customer service as measured by our Customer Satisfaction Survey. A large percentage of the total conservation targets are achieved through Key Account Customers. KAMs are given a specific target of GWh savings that need to be achieved through their customers. The customer care activities span across multiple disciplines within BC Hydro such as load interconnection, metering and billing, and negotiations and disputes.

2. Engage industry through a specific targeted program called the Power Smart Alliance.[14]

   The Power Smart Alliance is administered by BC Hydro and comprises a network of independent lighting, electrical, mechanical, and system professionals who assist BC Hydro customers in identifying, investigating, and implementing DSM and conservation solutions. There are over 600 firms who comprise the Power Smart Alliance. The core components of the Power Smart Alliance are day-to-day program support, customer referrals, supporting Alliance members to leverage the Power Smart brand, and developing and delivering technical information and training. Membership is free for all industry firms that meet eligibility requirements.

3. Motivate customers to conserve and be more energy efficient through investing in public awareness and education, with specific curriculum available for the education system.[15] The education component is important as students are future BC Hydro customers and it can be important to engrain behaviors early.

4. Engage stakeholders and First Nations through an external planning committee called the Electricity Conservation and Efficiency (EC&E) Committee.

   The EC&E Committee was formed in 2006 and has a membership of approximately 20 external people, representing a wide range of interests (regulatory intervener groups, environmental organizations, academia, provincial and local governments and First Nations). The mandate of the EC&E Committee is to provide informal advice and formal recommendations to BC Hydro on how to improve the design and delivery of current electricity conservation and efficiency programs and develop new programs and initiatives that will successfully engage British Columbians across the province in electricity conservation

---

[14]https://www.bchydro.com/powersmart/alliance.html.

[15]https://www.bchydro.com/community/youth_education.html.

and efficiency. An area where the EC&E Committee has contributed heavily includes the development of a strategic framework for planning for DSM, which led to the design of different resource options considered in BC Hydro's Integrated Resource Plan.

5. Consult with First Nations and the public through the Integrated Resource Planning process on the treatment of DSM within the overall resource plan.

   Consultation on BC Hydro's IRP typically occurs within a five-year planning cycle.[16] In the past, the IRP consultation process has targeted different streams for consultation, specifically, First Nations, the public, stakeholders, and a technical advisory committee. Consultation activities are scheduled around different phases of development of the IRP, including inputs, analysis, draft, and final plans.

   Questions explored include support for current levels of DSM investment as well as expanding efforts going forward. Feedback from the IRP consultation process has been important in reinforcing the level of DSM in BC Hydro's resource plan. In particular, participants in the IRP consultation process typically have strong or very strong levels of support for DSM activity overall.

6. Work with employees to conserve at their headquarters and "lead by example."

   Employees interestingly enough were one of the latter audiences to be targeted through the 25-year history of Power Smart. There was the view that in addition to being an employee, they were also BC Hydro customers; therefore, they would subscribe and participate like any other "knowledgeable" customer and take up the program offerings. It was actually some of our customers who made the connection and observed that BC Hydro was not really "walking the talk" on its own internal conservation. In addition, engagement through customer interactions revealed that program offerings had not been "tested" or "evaluated" from BC Hydro's experience in its own operations and capital improvement efforts.

   As a result, the *Lead by Example* concept was initiated in 2007 mostly to reflect external customer expectations on understanding the challenges and hurdles of program implementation, but also as a way for the company to become more efficient in running our own operations, saving money, and lowering our environmental impact. A key outcome of this effort was to tap the innovation and passion of employees and establish a base for testing and evaluating technical or process solutions that may in turn prove valuable for our own customers. Many employees, particularly those in a younger demographic, express the desire to make a difference in their workplace and link their personal values with those of the company. Green teams under the Lead by Example program spurred conservation workplace leadership, resulting in 350 members at 38 sites, ensuring that most employees have access to or leadership within their

---

[16]Per legislation, BC Hydro is required to submit an IRP to the BC government every five years. As part of its IRP submission to government, BC Hydro is also required by legislation to provide a description of the consultations on the development of the integrated resource plan.

workplace on this topic. Training, program sharing, research, and competitions were all facets of attraction to identify leaders to sustain the program. An annual environmental conservation leadership awards event also captures the results and innovation of individuals and teams, which is communicated across the entire workforce. The awards event used to be stand-alone, but now has been integrated as part of BC Hydro's Employee Awards, the highest recognition of employee leadership in support of our organizational values. In 2013, more than ten sustainability-related nominations were made and three of those ended up as winners or finalists.

The other aspect of the Lead by Example program was to identify hard-wired electricity savings across the operations and projects through efficiency audits, capital program evaluations, and procurement specifications that have led to business process embedment that will be sustained. Examples range from conducting promotional campaigns to keep the warehouse bay doors closed in garages during winter months, to following LEED building standards for renovations targeting a 30 % reduction in water and energy use, and the procurement of environmentally friendly products and materials in our operational supplies. BC Hydro's operations are regulated under provincial Carbon Neutral regulations aimed at reducing emissions in the public sector and these efforts also have the co-benefit of reducing emissions from fossil fuel electricity and fleet, translating to cost savings by avoiding required annual offset costs priced at $25/tonne of $CO_2e$.[17]

### *3.5.2 Generating Internal Acceptance*

One of the challenges for Power Smart in the early years was that it was met with some internal skepticism. Many BC Hydro employees outside of Power Smart did not understand demand-side management or the rationale for investing in such programs as they felt they could not be measured in the same way as a supply-side generation resource (e.g., metered GWh). This response is not surprising as many employees in BC Hydro had been involved with the significant hydroelectric generation development that had occurred in the decades prior in the province, and so for them, pursuing Power Smart was a culture shift from where BC Hydro had been.

One of the ways to get around this internal barrier was for Power Smart to increase the level of rigor in the information it collected on energy savings, and in particular, to try to describe the resource in a way that was comparable to a supply-side generation resource to facilitate comparisons within BC Hydro's integrated resource planning process—a catchphrase in the electric industry refers to this as "building a DSM power plant." Some of this was happening elsewhere

---

[17] http://www2.news.gov.bc.ca/news_releases_2013-2017/2014ENV0008-000250.html.

in the electric industry at the time, and so Power Smart was both able to draw from and apply more sophisticated practices around evaluation, measurement, and verification of energy savings, as well as tracking information on program performance to inform current and future offerings.

More recently, BC Hydro has developed a risk framework to characterize the uncertainty with energy savings to use in the long-term resource planning process. The risk framework uses a subjective probability elicitation approach: Internal program managers provide a range of uncertainty based on energy savings and participation rates. This information is used to develop a "bottom up" range of uncertainty, which is further reviewed and adjusted based on higher level considerations (interaction between DSM tools, macroeconomic drivers). Monte Carlo analysis is used to derive "P50," "P10," and "P90" estimates (a P50 estimate is a statistical confidence interval meaning that 50 % of the estimates exceed the P50 estimate and 50 % of the estimates will be lower), which are then used to inform base and contingency resource plans in long-term planning. In addition to meeting internal needs, this work is discussed before the BC Utilities Commission, the provincial regulator, and has helped build comfort with planned levels of DSM. Randy Reimann, Director of Energy Planning was a KAM during the start of Power Smart and then worked with BC Hydro's customers to understand their electricity needs, including opportunities to conserve or be more efficient.

> "When Power Smart started, there was a whirlwind of activity surrounding program development and targeting energy savings projects. I remember the phrase 'ready, fire, aim' used during that period because taking risks and making mistakes was important at that time so as not to limit opportunities. One thing that stands out to me now is how much we've improved the sophistication of the analysis and program development over the last 25 years. In particular, our risk assessment process for DSM has been very important in the resource planning process to be able to compare it to supply side resources and inform our contingency resource planning. I think it's been necessary to sustain the Power Smart program over this time and to justify it to internal and external customers and stakeholders."[18]

### 3.5.3 Getting Out of the Utility Mind-set

One way Power Smart diverged from "traditional" utility practice early on was to use innovative marketing techniques, even though they may be relatively modest relative to other companies' marketing budgets. Marketing and/or advertising expenditures are often highly scrutinized at a regulated utility such as BC Hydro. However, recognition that marketing activities are critical to deliver energy savings has grown internally over the years. The rationale for marketing is based on the need to change people's behavior through education and social influence. And there is also the realization that there is a highly competitive market for service

---

[18]Personal Communication—June 17, 2015.

providers (versus electricity providers). Power Smart program managers are now encouraged to compare themselves to other marketers in developing and implementing their programs, not necessarily to what other utilities are doing.

One of the marketing strategies used in Power Smart is "meeting customers where they are." This refers to developing approaches to engage with the customer based on their interests and actions, not just from the utility perspective. This has led to different strategies to rolling out programs such as engaging local sport celebrities (such as Steve Nash, recently retired two-time National Basketball Association Most Valuable Player, born in British Columbia) to participate in awareness around joining and pledging a 10 % reduction in residential consumption as part of campaigns and developing targeted partnerships with influential retailers. This program is known as Team Power Smart which looks to promote energy-efficient and conservation behaviors among residential customers. Team Power Smart has enlisted the help of a number of high profile business, political, community, and sports leaders who have roots in BC and share a passion for energy efficiency and conservation. Team Power Smart celebrity leaders have been a critical component of the "power of norms" strategy, where the influence these leaders have through their network or the media helps to elevate energy efficiency to a higher level of status. People tend to emulate those they admire and respect and are therefore, more likely to act upon conservation messages delivered by community leaders. These leaders continue to be committed to making changes in their own lives that lead to energy conservation and efficiency, with the goal of encouraging British Columbians to participate in conservation activities.

Another good example of an innovative marketing approach is Power Smart's compact fluorescent light bulb campaign that was initiated in the early 2000s. Whereas others in the industry were taking the approach to give away bulbs at community events or in mail outs, BC Hydro worked hard to establish relationships with large retailers to put the energy-efficient products on their shelves. BC Hydro did not see any long-term value in coming between a customer and their preferred retailer, nor to get into the business of retail product sales. These types of partnerships remain today and there are now over 1300 partners who are highly important in sustaining and leveraging the delivery of Power Smart programs. Jim Nelson, Conservation and Energy Management Senior Marketing Manager describes the multi-pronged approach to marketing: understanding customer behaviour, utilizing social science theory and using the power of strategic partnerships.

> "We use a social marketing framework within Power Smart based on a set of 10 principles or inspirations that were developed for Power Smart program managers to help deliver their DSM programs in innovative, effective ways. These inspirations are meant to be "tools in the toolkit" designed to stimulate ideas and leverage insights that can help catalyze change, rather than a set of rules that must be strictly followed. Some of the inspirations in the social marketing framework include: understand people, not customers; make an emotional connection to generate excitement and awareness; and remove acceptance issues."[19]

---

[19]Personal Communication—June 18, 2014.

### 3.5.4 Organizational Structure and the Importance of an Internal Champion

A common theme emerged during the interviews completed for this chapter: the importance of an organizational structure that sustains internal support and robust program management. In particular, having a representative on the executive team responsible for Power Smart was an advantage for the periods over the 25-year history—and conversely, a detriment for the periods where there was not an executive champion. The advantage of having an executive responsible for Power Smart is obvious–they can promote knowledge and understanding of the value proposition within the company's strategy, and they can serve as an advocate internally, for the customer.

Other organizational changes over the years have included Power Smart functioning as a separate business unit versus being rolled within a broader function. When Power Smart started, it was viewed as advantageous being a separate business unit. There were high goals to achieve new energy savings and operating independently allowed the new business unit to move quickly outside the often hierarchical utility structure. A disadvantage of this arrangement was that there was less knowledge transfer internally and consequently there was some internal skepticism over what this new group was delivering in the traditional context of utility services. As organizational changes occur, Power Smart is mindful of the strengths and weaknesses of different structures and seeks to play to the strengths and minimize weaknesses. Charles Reid, recently retired BC Hydro CEO has a unique perspective on DSM as he came from the forest industry where his former company was a BC Hydro customer and Power Smart participant. From his perspective as a BC Hydro customer, organizational stability was important.

> "When we interact with our customers be it on a billing issue or when they are needing help with a Power Smart business case for an asset upgrade, we as the utility need to work with them as a team to understand their business cycle, financial constraints, and operating environment - so we can add value along that entire relationship from the tactical issues to the strategic opportunities. If we constantly change direction in our programs or staff, it doesn't engender the necessary trust with the customer, nor support the utility's business objectives such as sustained DSM program savings."[20]

### 3.5.5 Government Policy Support

A final lesson from Power Smart's history is that having government policy specifying DSM targets has been important in embedding Power Smart as part of BC Hydro's core business. As noted the increased emphasis on DSM in energy policy and legislation sets the bar high on program delivery. This policy

---

[20]Personal Communication—September 18, 2014.

framework has been informed by ensuring the program results meet energy policy expectations, that co-benefits like customers' satisfaction and increased economic activity are understood, as well as participating in regulatory review of investments and savings. These all factor into creating the foundation for sustained direction and investment in DSM programs, as reflected on over the 20-year tenure by Steve Hobson, Director, Conservation and Energy Management.

> "This policy support has proved to be very important for BC Hydro to continue to invest in DSM. Just prior to the government's energy policy in 2008, I recall a conversation between Bev van Ruyven and our CEO at the time, Bob Elton. Bob offered some advice to us, which was this; 'You are going to have very few windows to embed DSM as a resource option and part of BC Hydro's business–and this is one of them.' Bev took a leave for six months to work with the government as they were revisiting their provincial Energy Plan and helped influence the inclusion of the 50 % DSM goal. This was a big goal, a stretch goal, but one that was very important to establish government and BC Hydro support for DSM as a resource with our customers and stakeholders."[21]

## 3.6 The Future for DSM

Looking ahead to the next decade, there is a good opportunity for BC Hydro's DSM programs to continue to build off the work and experience it has achieved over the past 25 years to deliver additional energy savings, but there are also new challenges to sustain DSM. As a cost-effective, low- or zero-carbon resource, there is increased interest in doing more with DSM from a policy perspective. Yet in some jurisdictions with a longer track record with DSM, the low hanging fruit has been picked and costs for additional actions have increased.

BC Hydro must be attuned to market trends and developments to continue to advance; and some areas being watched include the following:

- Policy advancements: In BC, the current policy target is for BC Hydro to meet 66 % of incremental load growth with DSM.[22] In the USA, parallel efforts are aimed at creating the policy platform and impetus for energy efficiency program adoption and implementation. At the federal level, President Obama announced in May 2014 to pursue an additional $2 billion dollar goal in Federal energy efficiency upgrades to Federal buildings over the next three years. The investment extends and expands the President's commitment to energy upgrades of Federal buildings using long-term energy savings to pay for up-front costs, at no net cost to taxpayers. This action aligns with the goal to encourage private and public sector leaders to join the Better Buildings Challenge and continue improving the efficiency of American commercial, institutional, and multi-family buildings and industrial plants by 20 % or more over 10 years. In

---

[21]Personal Communication—June 18, 2014.

[22]http://www.leg.bc.ca/39th2nd/amend/gov17-2.htm#section2.

combination with the initial commitment of $2 billion dollars in 2011, there is a total of $4 billion dollars in energy efficiency performance contracts in the Federal sector through 2016.

At the state level, there are regulatory efforts, such as the California Energy Commission paving the way to ensure appliance manufacturers must certify appliance efficiency data to the Energy Commission in order to comply with state law,[23] and its new 2013 Building Energy Efficiency Standards are forecast to save Californians $1.6 billion in energy costs over the next 30 years. The standards, adopted in May, are 25 % more energy efficient than previous standards for residential construction and 30 % better for non-residential construction.[24]

- Technology advancements: Information technology and distributed generation may bring new opportunities for energy savings, such as the adoption and penetration of net zero energy buildings. Through the installation of smart meters, utilities have access to more sophisticated information to understand how their customers use and save electricity. With these opportunities come new challenges as well. Utilities may or may not be well positioned to respond to these market developments. The aggressive goal of net zero energy buildings will displace more customer load, prompting some industry analysts to ask whether a "death spiral" is imminent and whether utilities can recover enough revenue through fixed charges to account for the loss of energy sales. Utilities may not have the appropriate resources to analyze "big data" and where their value proposition lies, and thus, there may be a role for partnerships with third parties to tap into this information.
- Economic conditions: One trend worth noting is the emergence of the "new normal" as it is known within the industry–electricity growth forecast at lower than previously planned levels (e.g., 1 %) suggesting investments in DSM and more modest economic growth. In jurisdictions where DSM acquisition levels are determined through the integrated resource planning process, this has the potential to lead to reduced need for new resource acquisitions for the industry overall and has the effect of lowering some of the conservation potential needed to meet aggressive policy targets.
- Increase in the need for new capacity resources: Over the past decade, BC Hydro's electricity needs diverged somewhat from its counterparts in North America. BC Hydro's electricity needs were greater for energy as its existing hydroelectric system had ample capacity to supply customer demand. However, customer demand growth has caught up to hydroelectric system capacity and BC Hydro now has a need for new capacity resources. Looking forward, BC Hydro will build off the experience of other utilities in North America to determine whether capacity-focused or demand response programs can be developed for the BC market.

---

[23] http://www.energy.ca.gov/appliances/. 2014.

[24] http://www.energy.ca.gov/efficiency/savings.html. 2012.

One way for BC Hydro to respond to potential future developments is to build off its strengths with engaging customers and broaden its strategies to tap into new energy savings opportunities. To date, many of the strategies to reach customers for new energy savings opportunities have been transaction-based: for example, replacing less efficient lighting or appliances. While this strategy has been effective, it still may miss additional energy savings opportunities. In particular, there is an opportunity to develop strategies around how the customers use electricity. Power Smart has explored this area, and one example is its Strategic Energy Management Program (SEMP) for its industrial customers. The program is based on providing funding for an energy manager position at some of BC Hydro's largest industrial customers who work "in-house" to identify and implement new energy savings opportunities. Because of where they are situated, energy managers' knowledge and understanding of ways to achieve viable energy savings exceed those of the utility. The industrial SEMP program has been successful, so BC Hydro has looked to replicate this program with its commercial customers. Expanding this type of approach, and the learnings to other customer segments may deepen the reach for savings.

Finally, there may be opportunities to facilitate nontraditional financing of energy efficiency investments, such as Green Revolving Funds, where the utility or another entity provides the upfront capital for an energy efficiency investment and applicants pay back the capital outlay based on project savings (e.g., reduced electricity bills). Such financing arrangements are anticipated to be helpful for customers who have limited access to capital. BC Hydro's recently retired CEO Charles Reid also reflects that there may be opportunities in the industrial sector to explore risk-sharing options on projects over the long term rather than under traditional financing mechanisms and payback periods. These programs have the potential to be tailored, innovative and further support the partnership model many utilities seek to grow market share and a strong sustained rate base in a cost-constrained competitive utility environment.

## 3.7 Conclusion

DSM offers opportunities to assess how to balance the triple bottom line approach to sustainability.

- Economic: The economic driver argument for DSM has proven out over time. DSM is a cost-effective resource option and provides economic benefits to customers through lowering electricity bills, making them more competitive through enhanced efficiency, as well as supporting customer green marketing opportunities. The economic drivers for DSM have changed over time though. They weaken during periods of low electricity market prices and they can be particularly challenged when the market/regulatory structure does not provide sufficient protection or support for the DSM investment. During these periods, it may be more important to lean more heavily on the other two pillars to rationalize the investment.

- Environmental: A key value driver for Power Smart has been the concept that avoiding new generation or transmission and distribution assets not only defers the costs associated with expending capital, but could avoid the resultant environmental impacts from the generation facilities and related linear development. So despite the economic driver being the foundation for the business case, adoption by customers was leveraged moreso through the environmental argument and positioning.
- Social: The one constant that has been maintained throughout the variations in the economic and environmental positioning and benefits realization over the past 25 years has been the social pillar. It originally was a marketing and branding exercise to create recognition of Power Smart programs as encouraging users to be "smart with their power." It has become more sophisticated and links behavior and culture change across all customer classes under the broader conservation umbrella opportunities. It has created the ability for innovation and testing of new program concepts and ideas, which has led to increasingly best in class program design and replicated results over 25 years of program delivery. There is also a community and moral suasion component, which is shifting the culture among all customers to not have to be incented to conserve, but that it makes business sense to invest and practice conservation behaviors. The moral suasion is also two way. BC Hydro is now expected to sustain the investment and program innovation and offerings to fulfill its 66 % DSM requirement and continue to meet customer expectations. More importantly, it is also increasingly viewed as the baseline for continued consent to operate on new required supply-side generation projects.

BC Hydro's experience with DSM over the last 25 years has been that it robustly achieves triple bottom line outcomes. It iterates the economic, environmental and social aspects of the program with continuous improvement and adaptation to changing drivers and expectations inside the utility framework and outside with customers and stakeholders. Overall, the ability to sustain Power Smart at BC Hydro over a 25-year period comes down to a few key things. Communicating the learnings, failures, and successes of our own efforts with our customers, employees, and stakeholders has created better partnerships and program design. Securing the support of key influencers—within BC Hydro, within government and within the community—has been immensely important for DSM sustainment overall. And finally, being able to "think outside the box" both in terms of marketing programs to BC Hydro customers and rationalizing the Power Smart investment within BC Hydro have enabled Power Smart to have longevity and for DSM to stay fresh and relevant.

# Chapter 4
# Consolidated Edison Company of New York: Distributed Generation

**Jim Skillman and Morgan Scott**

**Abstract** Sustainability is achieved when a company is able to balance its financial, natural, and social capital, leveraging them to derive value for its stakeholders. An appreciation of how these forces work together to positively influence both the development and success of a business, as well as the business' long-term societal goals, can mean the difference between simply surviving as a company, and thriving. The utility industry is often referred to as being on the verge of a revolution, as technology develops at a quicker pace and customers become more deeply engaged in their energy choices. Customers are evaluating their opportunities for on-site generation, as they now have the option to incorporate cost-effective, cleaner renewable technologies while potentially providing themselves a level of self-sufficiency and resiliency. Distributed generation (DG) can offer a truly sustainable solution—empowering customers with new energy sources, allowing for the development of cleaner technologies such as solar power and energy storage, and providing businesses with new opportunities to improve resiliency and defer or completely eliminate long-term infrastructure investment through targeted reductions in customer load. Con Edison is looking at how DG can be better integrated into its system planning process and long-term strategy. While the technology can present challenges to the utility business model and system design, Con Edison also acknowledges that DG brings benefits to its system and to its customers. Where Con Edison once developed energy forecasts and infrastructure plans based on the unidirectional flow of energy from large-scale generating facilities through to the customer meter, it now considers how a two-way flow of energy, changing load needs, and individual customer impact on the power system are driving the evolution of the modern utility.

---

J. Skillman (✉) · M. Scott
Con Edison, New York City, NY, USA
e-mail: SKILLMANJ@coned.com

## 4.1 Introduction

Sustainability is achieved when a company is able to balance its financial, natural, and social capital, leveraging them to derive value for its stakeholders. An appreciation of how these forces work together to positively influence both the development and success of a business, as well as the business' long-term societal goals, can mean the difference between simply surviving as a company, and thriving. The utility industry is often referred to as being on the verge of a revolution, as technology develops at a quicker pace and customers become more engaged in their energy choices. Customers are evaluating their opportunities for on-site generation, as they now have the option to incorporate cost-effective, cleaner renewable technologies while potentially providing themselves a level of self-sufficiency and resiliency. Distributed generation (DG) can offer a sustainable solution—empowering customers with new energy sources, allowing for development of cleaner technologies such as solar power and energy storage, and providing businesses with new opportunities to improve resiliency. For utilities, DG can help defer long-term infrastructure investment through targeted reductions in customer load.

At Con Edison, we seek to provide safe, reliable, and cost-effective energy services for all our customers. While many benefits can be realized through the integration of DG solutions, there have been lessons learned through the experiences gained with each new project. From technical challenges associated with project interconnection into a complicated underground network system to higher-level policy challenges related to net metering, DG is hardly an easy fix to the question of energy independence at the customer level. However, close relationships with our stakeholders—such as those with New York University (NYU), Jetro Cash and Carry, and the City of New York outlined in the case studies—help us to develop solutions that work for both Con Edison and our customers. Overall, the Company is looking at how DG can be better integrated into our system planning process and long-term strategy. Where Con Edison once developed energy forecasts and infrastructure plans based on the unidirectional flow of energy from large-scale generating facilities through the distribution system to the customer meter, we now consider how a two-way flow of energy, changing load needs, and individual customer impact on the power system are driving the evolution of the modern utility.

Con Edison's interest in DG is based in supporting our customer's choice of energy supply while capturing the opportunities presented by DG to improve upon our system reliability and resiliency. In the late 1990s, the New York Public Service Commission introduced revenue decoupling for New York's electric and gas services. Via decoupling, even when Con Edison delivers fewer kilowatt hours (kWh), or fewer kilowatts (kW) of demand, the Company may still earn its rate of return through surcharges. For this reason, the introduction of DG does not result in the same negative impact to the Company as it might for a vertically integrated utility and can, in fact, be a positive solution to certain system challenges. By reducing customer load overall and by shaving peak usage, Con Edison can reduce or even avoid significant capital expenditures altogether. This is good for both the company's and our customers' bottom line.

## 4.2 The Basics of Distributed Generation

Distributed generation is the installation of an electric generating system on customer premises that reduces the need to purchase energy from traditional energy providers, either the local utility or an energy service company (ESCO). Where customers were once simply an energy user receiving electricity flow one way from traditional generation sources, the installation of DG has changed their role into both an electric generation source and a consumer, resulting in two-way energy flow for the grid in some cases. The economic benefit of DG is derived either from harnessing a free source of renewable energy, such as solar or wind power, or through an increased overall efficiency of operation, primarily by using waste heat sources to drive a heating or absorption cooling cycle that would otherwise run on additional electrical equipment. Furthermore, the DG system may provide additional services to the customer, such as resiliency to storm-related outages or reactive power compensation, or to the utility, in the form of frequency regulation or demand reduction during peak electric usage periods. In some cases, these services are valuable enough to the utility or local Independent System Operator (ISO) to generate revenue streams that could reduce the payback period of participating DG systems.

In New York State, a distributed generation facility up to 20 MW in size that is dedicated to the support of nearby or associated load will be interconnected by the utility. For DG installations greater than 20 MW, the interconnection process follows the Federal Energy Regulatory Commission's Small-Generator Interconnection Procedures (FERC SGIP) under the jurisdiction of the New York ISO. For systems under 2 MW, the New York State Standardized Interconnection Requirements (NYS SIR) govern the process, expected time frame, and reporting requirements for DG interconnecting to utilities within New York State. The NYS SIR also addresses the technologies, customer classes, and size limits for installations eligible for net metering as per the latest NYS Public Service Commission orders and state law. Non-net meter eligible DG systems are typically served by Standby Service Rates, designed to recoup the utility's operational costs for installed electric system assets through fixed charges known as "Contract Demand." However, the Con Edison Electric Tariff contains specific carve-outs for technologies and capacities that are exempt from Standby Service Rates, for example, small installations less than fifteen percent of total load, or customers with less than 50 kW total demand. In addition, the Public Service Commission has specified certain "Designated Technologies" that are exempt from standby rates on the basis of being either a renewable energy technology or a small, highly efficient cogeneration system.

There are numerous types of DG installations throughout the Con Edison service territory, but the two primary technologies are combined heat and power (CHP) and solar photovoltaic (PV) generation. The cost to safely install these technologies without harming the grid is still relatively high; therefore, availability of subsidies and incentives can be critical factors in making installation affordable for customers. As with any new technology innovation, how the technology is treated within the utility rate and tariff structure plays a role in commercially viability. However, some regulatory approaches—like the full-retail net metering of PV systems used by New York State—can be economically beneficial for the DG customer, but shift cost to non-net-metered customers because net-metered customers do not pay their fair share to fund keeping the electrical system infrastructure reliable (wires, cable, base load power, etc.). Whether regulatory oversight can keep pace with rapid technology advancements in DG will directly impact the rate of adoption of DG solutions by customers and will determine whether future adoption benefits all customers or merely DG customers.

## 4.3 Combined Heat and Power

Con Edison's top source of distributed generation in terms of MW installed has been large CHP systems serving high-load customers. CHP, or cogeneration, systems are typically large-scale electric generator installations that capture the waste heat produced by the prime mover and generator to provide heat or hot water to the customer as well. The CHP system is typically run on natural gas, which can add an additional level of construction and fuel costs to add or upgrade existing gas services. Customers with existing gas service are required to first request a gas service review from their gas provider (within the Con Edison electric service territory that is either Con Edison or National Grid) to ensure that the existing service can handle the additional gas flow requirements of the planned DG installation. Con Edison's Gas Engineering will determine whether the existing service is sufficient to provide adequate pressure and flow for the DG installation, or whether a service upgrade or gas main extension will be required—at customer expense—which could be significant depending upon the distance to the nearest gas main. Customers are also required to work with the New York City Department of Buildings and Department of Environmental Protection or New York State's Department of Environmental Conservation to ensure necessary permits for equipment installation, and system emissions are secured—each of which are an expense. The effort is often worth the additional costs as CHP installations improve overall building efficiency and can reduce total energy costs significantly. Additionally, CHP can provide some customers with the possibility of "islanding," where the customer becomes completely self-reliant from the utility grid, providing a level of resiliency in the event of long-term disruptions to the electric distribution system.

Between 2006 and 2011, 90 MW of large (greater than 1 MW) CHP projects came online. While large project installation slowed in 2011, the market has shifted to many small capacity system installations. Currently, there are over 165 MW of CHP installed in the Con Edison territory, with multiple large-scale projects in development and an expected additional 100 MW of CHP expected to be installed by 2018. The small capacity CHP systems (less than 1 MW) have also experienced strong growth thanks to a new, modular CHP incentive program that was developed in conjunction with the New York State Energy Research and Development Authority (NYSERDA). These smaller systems tend to be sized to meet the building's thermal requirements, and, as such, have seen increased adoption rates as customers convert from #6 fuel oil to natural gas in order to meet NYC Department of Buildings and Department of Conservation requirements.

## 4.4 Photovoltaic Solar

Residential and commercial customers have also been adopting rooftop solar PV at an exponential rate (Fig. 4.1). Solar PV is an option for customers looking to integrate a renewable generation source with minimal operations and maintenance (O&M) costs. Even in NYC, where higher property values per square foot of interior space make other DG installations infeasible, PV, with the help of state incentives and city and federal tax breaks, has been able to find enough roof space to drive adoption. In the outer boroughs, where high-rise buildings make way for large warehouse properties, large PV installations have become the predominant choice for adoption of DG technology. However, the demographic driving the fastest growth has been the residential sector, with the availability of capital for third-party lease arrangements which allow the customer to adopt PV with little or no money down.

There are generally four primary criteria for customers to consider when determining whether a solar installation is appropriate:

- Is the location for the installation south facing?
- Is there roof shading from surrounding obstructions, such as trees and other buildings?
- Are there obstructions from other customer-owned rooftop equipment (HVAC systems, standpipes, chimneys, etc.)?
- Is the roof structurally sound with a remaining life span of no less than 20 years?

Additionally, Con Edison would advise customers to consider their electric usage patterns, personal usage behavior, and existing electric rate structure before considering PV as their first choice DG solution. For example, customers on monthly demand rates generally do not see much savings in their electric demand charges as the solar peak production either does not align with their usage peak or is not available during a peak usage time due to inconsistent production resulting from weather variations. For customers on an optional time-of-day rate, net-metered PV can defer high-cost kWh during the day, but the value of weekend PV production is a fraction of what it might be otherwise to the point where it might be beneficial to return to non-time-of-day rates. Residential customers, who generally have the most to gain with PV installations, should carefully consider third-party lease agreements that can eat away at the financial benefits a PV system can provide.

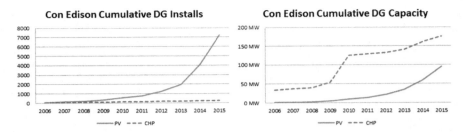

**Fig. 4.1** Growth of Distributed Generation at Con Edison

As is true of all DG installations, the customer will need to work with a third-party developer to design and install the solar array. This process will require two permits from the Department of Buildings—an electrical permit from the Bureau of Electrical Control and a building permit obtained by the third-party structural engineer. The installer will act on the customer's behalf to submit an interconnection request to Con Edison. Most PV systems under 2 MW will interconnect with the utility at little to no cost to the customer based upon current NYS SIR maximum limits. As systems get larger, installers should discuss their projects with the utility during the customer acquisition process because larger PV systems often incur costs beyond the SIR limits, such as utility system upgrades. In general, installers looking to build a system that is intended to export should contact the utility to ensure no adverse electric system conditions exist that would prevent PV system installation at little or no cost.

In the Con Edison service territory, customers have rapidly been installing solar PV panels. As of June 2015, 5500 customers have installed solar PV for a total capacity of 75.4 MW, growing from 34.8 MW at the end of 2013. Solar power production has quadrupled across our service territory in the last three years, with a compound annual growth rate of 75 %. Additionally, more than 60 MW of solar PV development is in the queue for interconnection. Large customer solar installations are on the rise thanks to a new NYSERDA incentive for commercial installations greater than 200 kW. In addition, innovations by Con Edison engineers now allow large PV systems to be installed on isolated network locations, allowing even more large installations to be completed.

## 4.5 Incentivizing Distributed Generation

### 4.5.1 *New York State Energy Research and Development Authority*

While most DG incentives in New York are managed by NYSERDA, Con Edison is responsible for assisting customers with the interconnection of their DG installations, but the Company has not historically directly incentivized customers.

We are currently exploring new incentives and arrangements in targeted areas where customer-sited reductions will help reduce infrastructure spending.

For CHP installations, NYSERDA has compiled a standard catalog of pre-approved and vetted CHP vendors and technologies that are incentivized as part of the CHP Acceleration Program with system capacities up to 1.3 MW. Additionally, incentives are available for installations greater than 1.3 MW that provide summer on-peak demand reduction. These incentives can reach up to 50 % of total project cost based on the performance of the system over a two-year period evaluating the energy generation (kWh), summer peak demand reduction (kW), and fuel conversion efficiency. In addition, Con Edison worked with NYSERDA to establish CHP Targeted Zones, indicating locations where CHP could assist in the offsetting of grid reinforcement or expansion. Projects developed within these zones are eligible for an additional ten percent incentive bonus.

NYSERDA also provides an array of incentives and financing assistance for solar installations. One example is the NY-Sun Megawatt Block Incentive Program. The goal of the program is to stimulate the solar installation market to increase the rate of customer installations while encouraging technological developments to lower installed system costs to a level at which the incentive is no longer needed. The incentive program assigns applicant PV systems an incentive value per watt until the combined capacity of those systems reach a specified MW limit, referred to as a "MW Block." The next applicants will receive a lower incentive as part of the succeeding MW Block until that specified MW limit is reached and so on until all funding is assigned. Using the Con Edison residential MW Block as an example, once 14 MW worth of projects are achieved (receiving an incentive of $1.00/W), that block is closed and the next projects in the queue—up to 6 MW—will receive $0.90/W. The process is continued until the total amount for the incentive program is reached—in this example, once 302 MW of residential solar installations in Con Edison's territory have been installed. Through the NY-Sun MW Block Incentive Program, a total 1680 MW of residential and non-residential installations have been incentivized across the state over the next 10 years, 605 MW of which fall within the Con Edison service territory (Table 4.1).

Similar to the CHP Targeted Zones, Con Edison worked with NYSERDA to establish Solar Empowerment Zones that correspond to networks that have traditional load relief needs and are also day-peaking. New installations in these zones receive a 25 % higher incentive than solar in non-targeted zones and may help defer future infrastructure investment.

## 4.5.2 Net Metering

No discussion of DG and incentives is complete without consideration of net metering. Net metering in New York State is defined as where the reverse flow of electricity is measured via the customer's meter so that the customer is credited

**Table 4.1** NY-Sun incentive blocks for residential solar in Con Edison's territory

| Block | MW | Incentive/W |
|---|---|---|
| 1 | 14 | $1.00 |
| 2 | 6 | $0.90 |
| 3 | 9 | $0.80 |
| 4 | 12 | $0.70 |
| 5 | 15 | $0.60 |
| 6 | 18 | $0.50 |
| 7 | 38 | $0.40 |
| 8 | 70 | $0.30 |
| 9 | 120 | $0.20 |
| Total | 302 | |

*Source* http://ny-sun.ny.gov/for-installers/megawatt-block-incentive-structure

at the retail rate for any electricity provided to the grid by the customer's generating equipment. Eligible technologies for net metering in New York State include solar, wind, fuel cells, and micro-CHP, which are small residential units up to 10 kW. For many customers, net metering will result in decreased monthly electric charges and is a driving factor in reducing the payback period for a DG installation. When customers produce more energy than they consume, the utility must credit the customer's account the following month. The New York Public Service Commission has set a cap on the amount of net metering generation in Con Edison's service territory, which is currently 33 MW of wind and 664 MW from all other net meter eligible technologies.

While DG offers utility benefits through the reduction of load on the system, that reduction can vary based on the technology deployed. Utilities such as Con Edison are required to provide reliable service to all customers, including those with DG installations that are net metered. In order to achieve that high level of reliability and service, we must continue to upgrade the electrical infrastructure. Under the current rate structure in New York, the required capital to fund these upgrades is recovered through tariffed rate mechanisms. When there is a reduction in revenue collection as a result of net metering, the concern, expressed by many throughout the electric industry, is that those customers who can afford DG installations avoid paying their fair share of the cost of broader system upgrades and maintenance, while those who cannot afford, or are not in a location where a DG installation is feasible, will shoulder an undue proportion of those costs (wires, lines, and other infrastructure). Ultimately, non-net-metered customers, who in many cases are the low-to-moderate income customers, subsidize typically wealthy or otherwise advantaged net-metered customers to the detriment of the system as a whole.

Con Edison's position is that as eligible technologies become competitive with conventional generation, net metering subsidies should end, and DG customers should be compensated for their contribution to, and pay their fair share for the benefits received from, the greater electric grid. These customers should assume their share of responsibility for the costs that their use of the electric system imposes on the system like any other customer.

## 4.6 The Con Edison Distributed Generation Process

Interconnection of DG to the electric distribution grid can have potentially negative effects on power quality and reliability of service when safety precautions and engineering controls are not in place. New York State has promulgated the general interconnection technical requirements in the NYS SIR which notes that the process for interconnecting a customer interested in a DG installation will vary based upon the maximum output capacity of the generator. A full description of the process can be found at the NYS Public Service Commission website; however, we will touch on a few of the finer points below to illustrate some of the complexities of installing DG within the Con Edison service territory. For more information on DG in NYC, Con Edison provides an online toolbox with DG information and links to helpful resources, which is easily accessible on the Con Edison DG Web site (conEd.com/dg).

The NYS SIR has two process paths for DG installations, an expedited application process for DG systems less than 50 kW and a standard process for those greater than 50 kW but less than 2 MW. However, for inverter-based systems less than 300 kW, which generally have a simpler interconnection process due to built-in UL1741 protections in the inverter, the SIR encourages utilities to use the expedited process. Within Con Edison's territory, due to the predominance of the network electric distribution system, even DG systems as small as 25 kW can cause unwanted cyclic operation of network protectors resulting in a reduction in network reliability. Therefore, as a general rule, all PV interconnection requests larger than 25 kW, as well as all non-PV DG requests, are reviewed by Con Edison's Distribution Engineering department to address the potential for unwarranted network protector cycling. For PV requests less than 25 kW, installers can self-certify that the installed system is completed and tested in accordance with NYS SIR standards. For all other requests, Con Edison will perform on-site inspections to verify proper operation.

For net-metered installations designed to be larger than 200 kW, Con Edison recommends a pre-screen be completed by the Con Edison Distributed Generation Ombudsperson's office. These reviews are critical, allowing for potential challenges to be identified upfront, educating customers early about those challenges and how they will impact the project timeline and costs. By doing this evaluation early in the customer acquisition process, the installer can avoid committing to project budgets before Con Edison Engineering identifies interconnection challenges further into project development. The pre-screen is initiated when a customer submits a letter of request which includes the location, anticipated size, account number associated with the installation, and a letter of authorization from the customer so that customer load information can be released to the installer.

The DG Ombudsperson will then pull the Electric Mains and Services map and identify the Con Edison infrastructure that serves the customer. If the customer is tied into the overhead system, there are no concerns about network protector cycling, and typically, any export can be handled by the same transformer that serves the customer. However, for those customers who are on an area network, there will need to be further evaluation to determine whether the network protectors in the area could be affected by the export of the net-metered DG system.

If there are potential issues, the pre-screen identifies that the project will need further engineering review and may incur interconnection costs to the customer. This allows the installer to inform the customer of the pre-screen findings of additional time and cost to their project, which allows the customer to make a more informed decision on whether or not to move forward with the project.

Once a customer agrees to move forward with a DG project, it is the installer's responsibility to design, contract for, and install the desired DG system. Any Con Edison system upgrades required, such as additional electric lines or gas services, will be handled in parallel to the customer work, and it is the customer's responsibility to pay for these upgrades. All inspections performed by Con Edison are free of charge, as is replacement of the standard meter with a net meter.

## 4.7 Case Studies

### 4.7.1 New York University Combined Heat and Power

There are numerous benefits to be garnered with a CHP system; however, the customer's needs will drive the technology type and capacity of the system installed. In the best-fit case, the facility should have consistent and preferably year-round thermal needs on-site with sufficient electric demand that can be offset by the CHP generator and footprint (Fig. 4.2). In NYC, much of the building stock does not fulfill all of these factors, which tends to reduce interest in these projects. When customers evaluate the significant initial cost of the installation, the ongoing O&M costs, combined with the opportunity cost associated with the loss of usable square footage (upward of 1000–3000 ft$^2$ for just a 1 MW generator, plus all associated additional piping, wiring, and ducting required),[1] CHP often does not provide enough benefit to justify the initial cost. As a result, the prime candidates for CHP installations tend to be large electric customers with significant thermal load, such as hospitals, university campuses, or large developments consisting of multiple mixed use buildings. So it was no surprise when NYU chose to repower and expand their already existing DG system to meet their full campus energy needs.

---

[1]As per the NYSERDA PON 2568 CHP Acceleration Program Attachment C—CHP System Catalog (revised 9/14) which lists pre-qualified COGEN units and their footprints upon installation. http://www.nyserda.ny.gov/-/media/Files/FO/Current%20Funding%20Opportunities/PON%20 2568/2568attc.pdf.

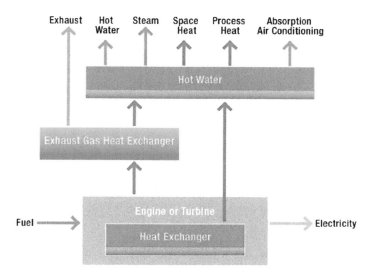

**Fig. 4.2** Typical cogeneration plant process. *Source* http://www.coned.com/dg/specs_tariffs/distributed-generation-guide-chp-customer-guide-version-2-november-2013.pdf

NYU had a number of drivers for the project including maintaining price stability, energy security, and improving environmental performance. Their Climate Action Plan, which guides the University's goals to reduce greenhouse gas (GHG) emissions, was also a contributing factor for this project, which today helps NYU avoid over 43,000 tons of $CO_2$ emissions annually. The system generates 13.4 MW of electricity and 90,000 pounds of steam hourly through the use of a steam turbine, two combustion turbines, and two heat recovery steam generators. This output meets the electrical needs for 22 campus buildings and steam and hot water needs for 37 campus buildings. The system also saves the campus approximately $5 million annually in energy costs.[2] "We've been engaged with NYU for over a decade on this project, and it's a prime example of how a utility and large-scale campus can come together to find an energy solution that works," says Michael Moccia, Customer Project Manager at Con Edison (Personal Communication, 2014).

While there were construction challenges on the side of the customer, the most significant challenge on the utility side of the project was the design of the services and the associated standby rates. To understand the full nature of the challenge, a discussion of the customer and utility perspective of standby rates is required. The basis for standby rates derives from how utilities recover costs to provide electric delivery service to customers, including the costs to install the customer's initial electric service connection to meet the maximum expected customer load. By collecting revenue from customers—in the form

---

[2] http://www.sourceone-energy.com/resources/case-studies/new-york-university-cogeneration-plant/.

of volumetric charges for energy consumed (kWh) and for peak demand (kW) required—the utility will be able to recover its costs for installing the customer's electric service as well as the costs for maintaining all upstream transmission and distribution infrastructure. So, from a utility perspective, when a DG facility is installed on an existing electric service that supplies the customer with energy that would otherwise be provided by Con Edison, there is a reduction to the effectiveness of the utility cost recovery mechanism. In order to prevent large DG owners from spreading the burden of cost recovery onto other rate payers, regulators and utilities instituted standby rates, which require the customer to pay fixed delivery charges for the full capability to deliver power through the local facilities. From a customer perspective, standby rates are often a matter of some contention. However, standby service from the utility gives customers the assurance that the utility will back up the DG facility should there be a planned or unplanned outage of the facility.

To complicate matters even further, NYU did not have a single electric service, but had many low tension, secondary services as well as transformer installation fed buildings. Instead of internally wiring the CHP facility to the individual buildings and removing the extra electric services from Con Edison, NYU determined they wanted to also keep the individual Con Edison services for redundancy. There was no appropriate service class for this type of arrangement, and an associated rate schedule needed to be negotiated. At the conclusion of negotiations, it was determined that NYU would be subject to the provisions of standby service for all accounts, and the generator would be allowed to connect in either parallel or isolated operation for resiliency. "We [Con Edison] really encountered challenges with the electrical services, but in the end, our close relationship with NYU allowed for a solution which both parties agreed to. It also exposed a potential pit fall in future projects of the same design that we can now better plan for going forward," says Moccia (Personal Communication, 2014).

Perhaps one of the most important aspects of this installation for NYU has been the ability to island from the grid, where the University's engineering staff is able to operate the main disconnect switches to isolate the incoming Con Edison services and allow the campus buildings to be served by the generator alone. This benefit was demonstrated in October 2012 during Superstorm Sandy. The storm caused outages to over one million customers in the Con Edison service territory, including areas of downtown Manhattan where the NYU campus is located. The campus was able to disconnect from Con Edison's electric service, and the CHP installation provided NYU with electricity, heat, and hot water. As a result, not only was the storm's impact reduced for those on campus, but the campus was able to be used by city officials to assist area residents who had been forced to evacuate their homes. The NYU CHP installation is a prime example of how a customer can achieve both environmental and reliability goals while also saving money and serving the community—truly sustainable.

## 4.7.2 Jetro Cash and Carry Solar Array

Jetro Cash and Carry (Jetro) is a grocery wholesaler serving supermarkets, grocery retailers, and foodservice operators. Jetro's location in the Hunts Point region of New York is known as a solar hot spot, with its large number of warehouses providing a vast array of flat roofs perfect for the installation of solar. While a large-scale solar project seemed ideal from a customer and developer standpoint, the interconnection became a significant challenge for Con Edison (Fig. 4.3).

The Jetro pre-screen process identified that the building had a high-tension isolated-network 460-V service. This type of service is characterized, in this case, by three different sets of high-voltage (13 kV) primary cable feeding local transformers which step down to the installation's service voltage of 460 V. Cable is routed from each transformer to an associated network protector which then connects to a common running bus, allowing any of the feeders to serve customer load at multiple takeoff points. The service is referred to as an isolated network because the transformers only provide power for this one customer installation due to the high demand of the original customer. Since then, the building has changed both ownership and business type, from manufacturing to a warehouse space, yet the installed utility infrastructure remained to serve the new customer. In some cases, these underutilized transformers are removed altogether, but in the case of Jetro, they were left in place to provide additional reliability to the customer.

The network protectors are designed to protect the public from extended duration high-voltage electric system faults. The Con Edison underground network consists of multiple independent high-voltage feeders coming into an area, stepping down to distribution voltage levels through distributed transformers, and tying the low-voltage cables together in a grid pattern. During normal operation, these multiple independent high-voltage feeds provide excellent reliability to the low-voltage network. In the event of a high-voltage feeder fault, the non-faulted feeders provide power to the grid and maintain power to the customer premise. Unfortunately, the grid also provides an energy source to back-feed the faulted feeder through its transformer connection, increasing the energy available to the fault and making it more dangerous to the general public. To eliminate this

**Fig. 4.3** View from the Jetro Cash and Carry rooftop. *Source* http://www.coned.com/newsroom/news/pr20131212.asp

second source of power to the fault, Con Edison installed thousands of network protectors throughout the underground networks to serve as automatic shutoff points in the event of back-feed toward a high-voltage feeder fault. The network protectors installed on the isolated network at Jetro act the same way, on a much smaller scale.

Unfortunately, the back-feed of power toward the network resulting from export of overproducing customer DG is interpreted by the network protectors as analogous conditions to a fault on the high-voltage system and will subsequently open the network protector automatically. The Jetro installation's planned 4760 high-efficiency panels would have a nameplate capacity of 1.56 MW, which would produce far more power than the building used. The export through the net meter would be directed toward the isolated network, and all of the interconnected network protectors would open simultaneously. The UL1741 protections of the inverter would see the loss of utility power and shut down to avoid the possibility of islanding, causing a complete outage to the building. Sensing the building had load to serve, the network protectors would automatically close again to provide power to that load, restoring the building after a momentary outage. Now that utility power is restored, the inverter would attempt to interconnect to the service after waiting the required five-minute stabilization period. After five minutes, the inverter interconnects automatically, the PV builds up the power provided until it exceeds building need and exports into the isolated network, and the network protectors open up again. This cycle of repeated momentary outages and five-minute power-ups would be repeated as long as the PV potential exceeded building load and would cause undue wear and tear on the utility's network protectors and customer's motors and electronic equipment.

As a result, the planned Jetro installation would have been rejected based upon the potential for network protector cycling and customer outages; however, Con Edison engineers designed a protection scheme, based upon smart grid technology developed for storm-hardening improvements, which would allow for limited export to back-feed through the network protectors while maintaining positive, remote operation of the network protectors. The protection scheme requires the installation of specially designed network protector relays that are programed to allow the network protector to remain closed until it detects a backflow of power from the customer equivalent to 50 % of the associated transformer's rating, at which point it would trip open automatically. To ensure a network protector could open on actual high-voltage feeder faults that resulted in backflow less than 50 % of transformer rating, a supervisory control and data acquisition (SCADA) system would be installed. This system allows for remote monitoring and operation of network protectors using cyber-secure wireless communication. In addition, a relay scheme typically utilized to prevent unintentional islanding of large CHP systems (as opposed to the intentional islanding that occurred at NYU following Superstorm Sandy) was incorporated to ensure that export capability was limited when one of the isolated network transformer's high-voltage feeder came out of service due to fault or routine maintenance. When the feeder comes out of service, the network protector relay sends a signal to the anti-islanding relay control box which in turn sends a signal for one of the customer's inverters to shut down, thereby reducing potential PV export to the levels acceptable for the remaining isolated network transformers.

According to Kwame Budu, Customer Project Manager at Con Edison, "this project is important because we were able to design a solution and change corporate policies to allow our customer to realize this renewable generation solution despite the challenges of being a 'high-tension' customer. This is a great innovation of PV installation to supplement the city's electrical usage and relieve Con Edison of a total of 1 MW of power on our grid" (Personal communication, 2014). The success of the Jetro project has translated to other projects as well. To date, six other installations have been successfully completed with the Smart Grid PV Pilot Program and another four installations are in progress. This project is an example of Con Edison's dedication to support customers interested in renewable generation. Instead of rejecting the project completely because of the risk it posed to the system, the company was able to use an innovative smart grid design to overcome the challenge to meet our customer and environmental goals. To date, Jetro has produced 1.8 million kWh of solar power and expects reduced direct emissions of 1212 tons of $CO_2e$ annually.[3]

### 4.7.3 New York City Wastewater Treatment Plant Solar Project

The solar PV installation on Port Richmond Wastewater Treatment Plant (WWTP) (Fig. 4.4) on Staten Island will have a total capacity of 1.27 MW, will produce over 2.3 million kWh annually, and reduce GHG emissions by 1642 metric tons $CO_2$.[4] The WWTP has a peak load of 3 MW and operates 24 hours a day, 7 days a week. NYC selected this site as one of the first public facilities to receive solar PV based upon its favorable rooftop—228,000 $ft^2$ of unobstructed flat roof space, with a newly installed roof membrane. The energy produced by the PV system will offset about ten percent of the total 17 million kWh the plant will use over the course of a year and is the largest of four solar installation projects NYC is undertaking in the Boroughs of Staten Island and the Bronx, the total of which will have an aggregate capacity of 1.85 MW.

These projects were implemented by NYC with two primary goals—first, to support NYC's sustainability goals outlined in *PlaNYC* and the *One City: Built to Last* policy documents, and, second, to evaluate the power purchase agreement (PPA) financing structure as a means to procuring more cost-effective solar installations. *PlaNYC*, launched in 2007, established a vision and a plan for making

---

[3]http://www.nydailynews.com/new-york/bronx/hunts-point-home-largest-solar-panelinstallation-nyc-article-1.1700460.

[4]http://www.nyc.gov/html/planyc/downloads/pdf/publications/NYC_GHG_Inventory_2013.pdf.

NYC a more sustainable city in consideration of a forecasted population of 9.1 million residents by 2030. The plan addressed a variety of concerns, including increasing access to public parks and green space, protecting the city's waterways, and improving air quality. In September 2014, Mayor de Blasio announced *One City: Built to Last*, an expansion of *PlaNYC* that sets NYC on a path to achieving an 80 % reduction in GHG emissions by 2050 and targets the building sector for significant carbon reductions toward that goal. Under the plan, public buildings will serve as models for sustainability. To achieve the *One City* goal, NYC intends to employ multiple strategies, including the follwing:

1. Investing in high-value projects;
2. Expanding solar on NYC rooftops;
3. Implementing deep retrofits in key facilities;
4. Improving building O&M;
5. Piloting new clean energy technologies.

Under the goal to expand their use of renewable technologies, NYC will install 100 MW of solar over the next ten years. The WWTP project was an opportunity for NYC to demonstrate using the PPA financing structure on NYC property and to develop best practices for procuring solar PV via a PPA structure that is envisioned to be used to meet the 100 MW goal. While the solar installation is located on NYC property and the electricity is used by the treatment plant, the installation itself is owned and operated by an energy service provider—Con Edison Solutions.

**Fig. 4.4** View from the Port Richmond WWTP rooftop. *Source* Port Richmond Waste Water Treatment Plant Solar Array, New York.Personal photograph by author. 2014

Michael Perna, vice president of Marketing and Business Development at Con Edison Solutions noted, "We are delighted to work with the City of New York on this important project and believe it is a model for expanding the installation of solar projects throughout the City."[5]

The project was initiated, in part, to assess the cost of power available under a PPA structure, to help NYC identify cost-effective approaches to expand renewable energy development in NYC, and to develop procedures allowing NYC to better plan for future PV installations. The economics of solar projects for NYC are a particular challenge because NYC buildings purchase their power at a favorable rate through the New York Power Authority (NYPA). In this project, by purchasing the solar power through a PPA, NYC avoids costs associated with initial project construction and ongoing maintenance. The PPA is structured in a way that encourages the solar developer to move efficiently through the design, permitting, and construction processes, as no payment is issued until the system is permitted, interconnected, and generating electricity. Furthermore, because the solar developer recoups its costs through monthly electricity payments made by NYC for solar electricity generated, the developer has an incentive to optimally maintain the PV system to ensure maximum output over the twenty-year term of the PPA.

As can be expected, the project has not been without its challenges. While a typical installation may take a year to complete, this project has taken more than two years to achieve final interconnection, totaling eight years from planning to interconnection. Because the PPA procurement structure was new to NYC, it took longer than normal to navigate the contracting process. However, now a replicable model has been created and will be used for larger-scale PPAs in support of NYC's 100 MW goal.

The complexity of the NYC permitting process also added some difficulty and delay to this project. Due to the sizes and complexities of the existing electric service equipment, the electrical permits were subject to review and were approved by the NYC Department of Buildings Electrical Advisory Board. The Advisory Board meets once a month, and depending on the volume of applications, some applications were delayed to the next meeting, which impacted the schedule more than once during the project.

From a technical standpoint, the electrical infrastructure of the WWTP added complexity to the project as well. Due to the plant's High Tension service (connected at 33 kV) and its proximity to Con Edison's medium voltage substation, the technical review process by Con Edison Distribution Engineering was extensive. Unlike most other DG interconnection processes, interconnecting directly into Con Edison's High Tension service in Staten Island requires a more in-depth engineering design review, utilization of equipment limited to Con Edison-approved manufacturers, and more detailed system and equipment testing procedures to be provided or approved by Con Edison. This resulted in frequent documentation reviews between the City and Con Edison, which caused further project delays.

---

[5]Personal Communication, 2014

Despite the challenges, this project is an excellent learning opportunity for all parties and will help inform NYC's approach to scaling up its development of renewable energy. NYC and the PPA provider, Con Edison Solutions, will work together to develop education-based programs throughout the community to spread awareness of renewable energy. As the community becomes versed in the benefits of solar generation, and more customers become interested in developing their own installations, Con Edison will be ready to help them meet their energy needs and integrate their installations onto the grid.

## 4.8 Distributed Generation and Con Edison's Sustainability Strategy

The customer, environmental, and system benefits of DG are clear from the three case studies described. These benefits were important factors in Con Edison's decision to facilitate increased DG penetration throughout the service territory as a part of a long-range sustainability strategy. The sustainability strategy was first established by the Company's Corporate Leadership Team in 2008. The principles and initiatives covered both the regulated and competitive businesses, with a focus on minimizing GHG emissions as an approach to mitigating climate change. At the time, the company was looking at ways to reduce Scope 1 GHG emissions as described in the American Clean Energy and Security Act (ACES Act)—more commonly known as the Waxman-Markey Bill—which also provided for the development of a federal cap-and-trade system for GHG emissions. The initial strategy consisted of the packaged set of initiatives that would lead to the achievement of the Scope 1 GHG reductions and defined how Con Edison would provide our customers and the public with efficient, clean, and sustainable energy. To do this, the company defined six principles stating that Con Edison would:

1. Model green behavior internally;
2. Promote green behavior to external stakeholders;
3. Innovate to meet customer preferences for a green lifestyle;
4. Partner with governments to shape policies and standards consistent with its sustainability vision;
5. Develop infrastructure to advance the use and delivery of value-creating clean energy alternatives;
6. Incorporate environmental and societal values in its decision making.[6]

These principles were supported by 35 initiatives and an overarching internal goal to reduce the company's direct GHG emissions by 40 % from the 2005 baseline by 2020. The 35 initiatives encompassed projects such as the addition of cleaner burning natural gas at Con Edison of New York's steam plants, the company's green fleet strategy, and efforts to reduce energy consumption at company office

---

[6]http://www.conedison.com/documents/2009-EHS-Report.pdf.

facilities. While the primary focus was on direct emissions, the strategy also incorporated initiatives that would help mitigate our customers' carbon footprints including our energy efficiency programs, converting customers from oil to natural gas, and promoting renewable generation through a focus on DG interconnection. For the purposes of the sustainability strategy, the focus of the DG initiative was its ability to help spur further development of renewable generation, namely rooftop solar.

The sustainability strategy served well for encouraging the development and communication of these GHG-reducing initiatives and allowed the company to achieve its 2020 goal nine years ahead of schedule in 2011.[7] Still, in 2013, the strategy was updated to broaden its environmental focus to better balance the values of business, environment, and community—commonly referred to as the triple bottom line. Con Edison's sustainability vision was also broadened, stating that, "we strive to be an enterprise valued by all stakeholders, incorporating a focus on business, environmental, and community value as the foundation of our company strategy and our employees' day-to-day activities."[8] Similar to the initial strategy, the revised version is built upon six principles

1. **Business Value**

    (a) Provide our customers and the public with safe, reliable, cost-effective, efficient, and clean energy;
    (b) Adapt our business model to industry changes and manage risk to meet our shareholders' expectation of financial stability and a growing dividend.

2. **Environment**

    (a) Reduce our environmental footprint, making wise and effective use of natural resources;
    (b) Address the challenges of climate change and its impact on our business.

3. **Community**

    (a) Develop a work environment that fosters a diverse, high-quality, and engaged workforce valuing safety, sustainability, and business ethics;
    (b) Support the vitality of our communities and engage our stakeholders regarding strategy and performance.

In addition to these principles, the company undertook an internal materiality assessment and identified 15 key sustainability aspects as being critical to our sustainability and long-term success. Figure 4.5 illustrates those areas which were identified as key sustainability aspects in 2013.

With the evolution of the strategy, the importance of the company's DG work became even greater. Instead of being recognized primarily as an opportunity to

---

[7] http://www.conedison.com/ehs/2011annualreport/environmentalstewardship/reducing-greenhouse-gases.html.

[8] http://conedison.com/ehs/2013-sustainability-report/1d-strategy.html.

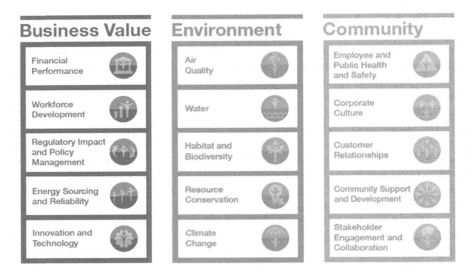

**Fig. 4.5** Con Edison's sustainability strategy

encourage renewable energy and reduce GHG emissions, it expanded to include planning and forecasting, system reliability, and helping our customers to manage their energy needs cost-effectively. Our efforts to assist our customers with DG integration will continue to be a focus—and is becoming an even larger consideration as we evaluate our long-term infrastructure needs.

## 4.9 Looking Forward—Utility of the Future

As we look toward the future at Con Edison, it is becoming apparent that distributed generation will have a greater presence in our service territory and will have an impact on our daily operations and future planning. The interest in DG from our customers, coupled with the incentives that are available through NYSERDA, has been driving adoption significantly over the past five years. The New York State Public Service Commission has begun taking bold steps to encourage the development of DG within the electric power industry. At the end of 2013, the Public Service Commission publicly identified five policy objectives that would drive their regulation of the distribution system for the coming years:

1. Empowering customers;
2. Leveraging customer contributions;
3. System-wide efficiency;
4. Fuel and resource diversity;
5. System reliability and resiliency.[9]

---

[9]http://apps.cio.ny.gov/apps/mediaContact/public/view.cfm?parm=7FFE507A-5056-9D2A-1050963802724A72.

This was followed in April 2014 with the release of the Public Service Commission Reforming the Energy Vision (REV) proceeding, outlining these policy objectives in more detail and adding a sixth objective of reducing carbon emissions.[10] Included in the proceeding was a discussion of what changes need to be made regarding market design to encourage the development and widespread deployment of distributed energy resources and energy efficiency solutions. Con Edison worked with the other utilities in the state to respond to the proceeding, and we agree that, with respect to the rapid pace of development and deployment of DG technologies, it is the right time to engage in a dialogue about how best to adapt the electric distribution system and the energy marketplace to promote DG integration while ensuring reliability. It is also Con Edison's position that this is an important time to further address the challenges of net metering and to better define how it should be utilized moving forward.

One of the major challenges of REV, as well as with DG adoption all around the world, is to determine the true value of the utility grid to the customer. While the falling cost of solar PV coupled with improvements in energy storage technology and the affordability of portable emergency generators may make the idea of customers going "off-grid" seem like an imminent threat, the reality is that most customers choose to stay connected to their utility. Customers who install solar PV find that net metering their excess generation is preferable to attempting to size a battery that can store that energy, and with the current NYS policy of retail value for solar production, it is more financially attractive as well. Often, customers are not able to generate at a capacity that meets their energy needs, and the variability of solar generation compounds this problem. The grid ensures customer reliability in this case, providing energy when their DG installation is unable to meet their electricity needs. The grid also provides customers voltage quality and start-up power, which for certain appliances such as air conditioners may require more power to start than a typical PV installation can provide.

In addition, while the grid was never intended as a two-way power flow platform, many utilities, including Con Edison, are finding ways to allow customers to utilize the grid as an asset. There are numerous proposals to build micro-grids within the Con Edison territory, where customer-sited DG would power multiple buildings in a geographic area to increase resiliency to utility outages from severe weather. The NY Prize Community Micro-grid Competition is providing NYSERDA funding for feasibility studies of micro-grids that incorporate a critical facility, like a hospital or police station. In addition, the recently enacted Community DG Net Metering Order looks to link customers who traditionally could not install solar (apartment dwellers and low-to-moderate income customers) with developers of large-scale PV systems. These programs and policies would not be possible if the grid did not offer an intrinsic value to its customers, far beyond that of just a simple conduit of electricity. The future of the utility grid holds myriad possibilities to expand that value to customers.

---

[10]State of New York Public Service Commission Case 14-M-0101, Order Instituting Proceeding(April 25, 2014)

In fact, Con Edison has already taken steps to explore how DG can be used to enhance our approach to system challenges. Three networks in the Borough of Brooklyn, served by two substations, Brownsville No. 1 and Brownsville No. 2, have seen significant growth and they are close to reaching maximum load capacity in the area. As of July 2015, forecasted summer overloads in this area of Brooklyn are 2016 (18 MW overload), 2017 (49 MW overload), and 2018 (58 MW overload). Traditionally, Con Edison would address this issue by building a new area substation. This would cost in the area of $1B, a cost that would ultimately be borne by our customers. Instead, in July 2014, the company released a request for information—to be followed by a request for proposal—asking qualified respondents to offer demand side management solutions which might include customer-sited and utility-sited resources, energy efficiency investments, distributed generation, and/or demand reduction programs. This is the first time that Con Edison has reached out to address a system constraint problem in this way and is a sign of how other utilities may begin to consider addressing their own infrastructure investment challenges in the future.

Distributed generation will be an influential factor in the evolution of the utility industry. While the technology can present challenges to the utility business model and system design, Con Edison acknowledges that DG can bring benefit to our system and our customers. We are committed to assisting our customers with their energy choices, whether it is incorporating energy efficient solutions, switching from heavy fuel oil to cleaner burning natural gas for heating, purchasing green power from ESCOs, or integrating DG onto the grid. Our customers are at the core of what we do, and our sustainability strategy will guide our transition into a utility of the future.

# Chapter 5
# CPS Energy: Clean Energy, Community, and Business Vitality

Monika Maeckle, Lisa Clyde and Kim Stoker

**Abstract** Like every utility, CPS Energy has grappled with balancing the need to satisfy increasing state and federal regulations with keeping rates low and service reliable. CPS Energy saw the writing on the wall regarding two older coal units that supplied approximately 20 % of the utility's total electricity. The challenge, however, became how to continue supplying reliable and affordable electric service to an economically disadvantaged customer base under increasing environmental regulation. In a creative turn, the company decided to leverage its substantial buying power to stimulate the local economy, create new jobs, attract world-class energy companies, increase renewable energy generation, and cultivate a more educated future workforce for San Antonio. The effort was called The New Energy Economy. As of October, 2015, CPS Energy's New Energy Economy has resulted in more than 842 new, full-time jobs, $52 million in payroll, $200 million in capital investment, and $4.2 million in education contributions.

Since 2010, the Environmental Protection Agency has moved to impose ever-stricter regulations on utilities regarding air pollution. Carbon, nitrogen oxides, mercury, and other pollutants have increasingly become subject to air quality testing standards, raising the bar for all entities that produce electricity for their communities to do so in a responsible way.

---

M. Maeckle
The Arsenal Group, San Antonio, USA

L. Clyde (✉)
Air Quality Compliance Engineer, CPS Energy, San Antonio, USA
e-mail: MAClyde@cpsenergy.com

K. Stoker
Environmental and Sustainability Director, CPS Energy, San Antonio, USA

## 5.1 Introduction

CPS Energy saw the writing on the wall regarding two older coal units that supplied approximately 20 % of the utility's total electricity. Could these aging units remain viable under increased regulation? The challenge became how to continue supplying reliable and affordable electric service to an economically disadvantaged customer base under increasing environmental regulation.

Would the utility spend billions of dollars to add controls and expensive upgrades to aging coal plants? Or would they invest in a more diverse, clean energy portfolio?

After thorough evaluation, CPS Energy chose the latter. In a creative turn, the company decided to woo "new energy" companies and partners to San Antonio by leveraging its substantial buying power to stimulate the local economy, create new, well-paid jobs, increase its renewable energy portfolio, and cultivate a more educated future workforce for San Antonio. The effort was called The New Energy Economy.

## 5.2 Birth of San Antonio's New Energy Economy

Less than a year after his arrival at CPS Energy in 2010, CEO Doyle N. Beneby made a prescient call: Shut down two coal units 15 years earlier than planned. The money that could have been spent retrofitting the Deely units to satisfy escalating environmental regulatory requirements would instead be redirected to develop cleaner energy sources and increase customer efficiency, while also creating jobs and local infrastructure.

"We're estimating that we are avoiding $3 billion of retrofits for a 33-year-old coal plant," Beneby told media in June of 2011.[1] "We've chosen to invest those dollars today in clean sources that are affordable," said Beneby. The utility would also mitigate risk through a more diversified energy portfolio. The announcement was unprecedented.

Four years later, CPS Energy's plan to directly leverage its considerable buying power on behalf of a new energy economy and its myriad educational, environmental, and economic benefits was recognized by the Pew Charitable Trusts[2] as, "leading the charge, establishing aggressive renewable energy goals."[3]

To kick-start the effort, Beneby announced six clean energy companies would relocate to San Antonio, bringing good-paying jobs, solar component manufacturing factories and infrastructure, and educational investment to the community. The New Energy Economy Initiative was born.

---

[1] http://insideclimatenews.org/news/20110621/cps-energy-shut-first-coal-plant-texas-solar-wind-epa.

[2] http://www.pewtrusts.org/en/research-and-analysis/issue-briefs/2014/11/texas-winds-generate-economic-growth.

[3] "Texas Winds Generate Economic Growth; clean economy rising" Pew Charitable Trusts. November 17, 2014.

## 5.3 What Is the New Energy Economy?

The New Energy Economy (NEE) is the suite of business, community, and environmental benefits of investing in technology innovations related to "new energy"—that is, more efficient and cleaner ways of providing power, such as solar, wind, coal with carbon capture, and new natural gas. It also includes efficiency programs such as demand response, which is tapping energy sources at nonpeak times to take advantage of better pricing on the spot energy market.

These new energy options contrast starkly with high carbon emitting sources such as un-scrubbed coal, older, less-efficient gas, and oil. Apart from cutting emissions and helping power suppliers satisfy increasingly stringent environmental requirements, new energy sources and their accompanying technologies serve as economic boons to communities that embrace them by creating a triple win—for the community, CPS Energy, and the participating NEE partners.

## 5.4 Background: San Antonio, a Unique Situation

Several factors conspired to allow the NEE to happen in San Antonio, Texas.

First, CPS Energy's unique corporate structure is cast not as a municipal department of the city, but an entity governed by a separate Board of Directors. This relative independence allowed the utility to tap community resources and its substantial buying power to aggressively recruit new energy companies to San Antonio. CPS Energy offered them guaranteed business; the companies agreed to bring jobs, economic, and educational investment if the companies brought clean sources of energy. Such private sector thinking is unusual in the world of municipal utilities.

In addition, a primary environmental and community concern for San Antonio is water. Until recently, the city held the dubious distinction of receiving the majority of its municipal water supply from the pristine Edwards Aquifer. After the historic 1950s drought, CPS Energy built two lakes, Calaveras and Braunig Lake, to provide cooling water for nearby power plants and was one of the first utilities in the nation to use treated wastewater to fill the reservoirs. With the reality of a changing, typically dry climate subject to frequent drought, the role of water in local electricity production has become magnified in recent years. And since coal plants require billions of gallons of water annually to create electricity, weaning the utility from coal not only would decrease emissions, but also would save water—not inconsequential for one of the country's fastest growing cities (Fig. 5.1).

San Antonio's water-challenged environment coupled with its status as one of the fastest growing but poorer cities also created a mindset for the utility to think differently. San Antonio is the seventh largest city in the USA and the third largest metropolitan area in Texas, behind Houston and Dallas-Ft. Worth. Yet unlike its

**Fig. 5.1** Calaveras Lake, south of San Antonio, TX

sister Texas cities, per capita household income hovers under $23,000 and median household income for the San Antonio Standard Metropolitan Statistical Area (SMSA) is $45,000. Twenty-four percent of San Antonio adults over age 25 have a college degree, and 46 % of households speak a language other than English at home. Further, 20 %[4] of San Antonio's population live below the poverty level; thus; affordability of utilities is very important.

Perceptions that San Antonio could not afford "expensive" renewable power or innovative technology needed to change. In addition, despite a low cost of living, inexpensive land, and an excellent rail network, clean-tech companies looking to relocate often overlooked San Antonio. Both new businesses and CPS Energy customers had to be convinced that:

1. San Antonio could support high-technology businesses, and
2. Clean, efficient power can be affordable.

In 2005, CPS Energy developed a vision that paved the way for future strategic planning, including "long-term financial and strategic resource planning; municipal, state, and federal legislative policy; motivating and providing clear direction for CPS Energy employees; and development of senior management accountabilities."[5]

As carbon emission concerns escalated and San Antonio and its electric load continued to grow, the likelihood of additional environmental regulations became evident. CPS Energy responded by developing the Save for Tomorrow Energy Plan (STEP) in 2008 with a goal of reducing demand by 771 MW by 2020, and a commitment to increase renewable generation to 1500 MW, or 20 % of total capacity, by 2020.

With the arrival of Beneby in 2010, the City of San Antonio and CPS Energy began an aggressive collaboration to address the challenge: How to create reliable, affordable electricity for a growing city with limited water resources while

---

[4]United States Census Bureau—State and County QuickFacts.
[5]CPS Energy Vision 2020 plan.

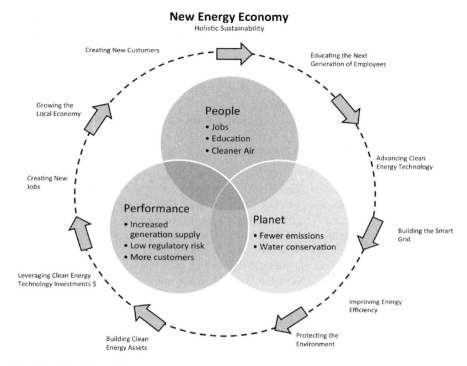

**Fig. 5.2** New Energy Economy

reducing emissions and staying in compliance with increasing environmental regulations in a deregulated energy market? The path CPS Energy walked led to the NEE.

## 5.5 Getting Started

In light of the growing uncertainties with coal generation, CPS Energy was already looking to further diversify the company's generation mix. With the arrival of the new CEO, a goal was set to achieve 20 % of generation from renewable energy and 65 % low- and no-carbon generation by 2020. These visions and concepts aligned closely with then Mayor Julián Castro and the City of San Antonio leadership's SA2020 initiative[6] (Fig. 5.2).

---

[6]http://www.sa2020.org/causes/environmental-sustainability/.

The City of San Antonio and CPS Energy made clear the NEE's goals:

- Rebalance CPS Energy's portfolio by achieving 1500 MW, or 20 % of generation capacity from renewable energy with a total of 65 % low- or no-carbon emitting power plants by 2020.
- Leverage utility investments, clean energy, and innovative technologies for job creation and education investment.
- Reduce emissions by an amount equal to removing almost 1.5 million cars from the road.
- Fuel investment in the economy and education of San Antonio.

Cross-functional teams were formed at CPS including research and development, finance, generation, distribution, environmental, business development, and legal. The teams began evaluating various technologies, developing business cases for clean generation sources in terms of value to the utility, customers, and the environment. Requests for Proposals (RFPs) were issued with the technical specifications defined by the cross-functional teams, as well as requirements for economic development. Respondents were encouraged to propose economic development packages that included local offices, headquarters in San Antonio, job headcounts, various types of jobs, defining minimum and average salaries, capital investment, direct education investments (pre-K to post-graduate), scholarships, internships, and more.

Prospects for business partnerships were identified using technology data, strategic fit with company goals, regulatory direction, and economic development models. Relationships with other utilities, research groups, consultants, and specialists were heavily utilized to evaluate potential technologies, vendors and NEE partners.

Stoplight charts were used at various intervals to give visual clarity. Weights were assigned to each category selected as important for a given project and technology. Values were assigned to the various stoplight colors in the weighted categories to determine the ranking of candidates. A sample stoplight chart is shown in Fig. 5.3.

Simultaneously, CPS Energy employees were educated about NEE through town hall discussions and annual employee development conferences. It became part of the vocabulary in every staff discussion, and a driving force behind the development of each business unit's strategic plan. CPS Energy took every opportunity to highlight the strategy, the objectives, and the vision and made state, local, and national news with announcements for every request for proposal (RFP) that went out, and agreement reached (Fig. 5.4).

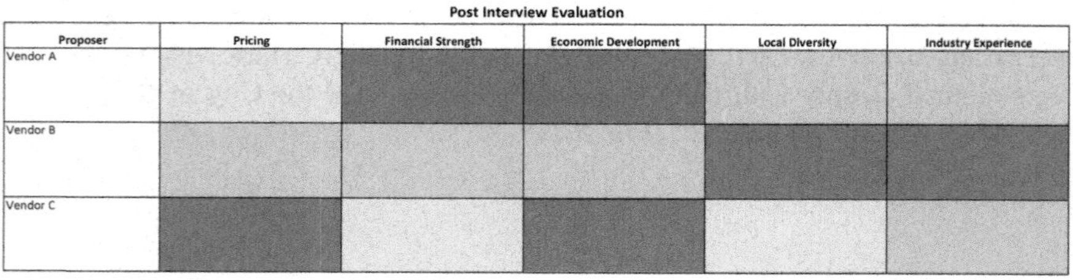

**Fig. 5.3** Post-interview evaluation

Fig. 5.4 The Texas Tribune

"The city's plan—spurred by a unique partnership between CPS Energy, its municipal utility and a solar power provider—has quietly brought in hundreds of jobs and millions of dollars in investment. It is enabling San Antonio to transition away from coal-fired power generation and helping Texas become a national leader in the industry."—The Texas Tribune, Sept. 30, 2014[7,8]

## 5.6 Stakeholder Engagement

The NEE is inherently a win–win proposition, from jobs and economic development to additional electric generation and cleaner air—but CPS Energy decided to take it a step further by supporting schools.

To support local educational initiatives, train the next innovative business owners, and develop the next generation of skilled CPS Energy employees, the NEE agreements specifically required that education investments be in the curricula of science, technology, engineering, and math (STEM). These fields are directly related to the generation of electricity and establish an ongoing investment in cultivating a future workforce. They also promote innovation and healthy living.

---

[7]The Texas Tribune is a nonpartisan, nonprofit media organization that informs Texans—and engages with them—about public policy, politics, government, and statewide issues.

[8]http://www.texastribune.org/2014/09/30/texas-only-solar-panel-manufacturer-ramps-producti/.

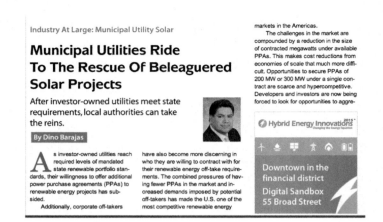

Fig. 5.5 Solar Industry

Former Mayor Castro championed the initiative and embraced it as part of his overall SA2020[9] strategic plan. Environmental groups such as the Sierra Club became advocates, recognizing CPS Energy for flipping the switch on the first utility scale solar project in the state.[10] Educational organizations praised the inclusion of STEM funds for elementary, secondary schools, colleges, and universities. CPS Energy customers appreciated educational investments, and everyone supported the clean-tech job development potential. CPS Energy employees volunteered for project teams to carry out NEE initiatives, providing opportunities for employee growth and a renewed feeling of involvement. Stakeholder buy-in was powerful and critical to success (Fig. 5.5).

> "The industry has already seen as a success story utilizing the sliding scale economic development model with the City of San Antonio's Alamo projects.—Solar Industry Magazine, August, 2014"[11]

When the nondisclosure agreements insisted upon by partner companies for competitive reasons were challenged by local media, CPS Energy was able to counter by reporting nonconfidential facts in the company's blog, *Energized*.

"It's competitive," said Analyst Matthew Feinstein of Lux Research in a July 10, 2013, blog post that addressed criticism of the NEE solar purchase power agreement.[12] "If I'm First Solar, I don't want a utility in Massachusetts to know what I'm

---

[9]SA2020 is a community vision for the future of San Antonio. It is a list of goals created by the people of San Antonio in 2010 based on their collective vision for our city in the year 2020; http://www.sa2020.org/what-is-sa2020/.

[10]"New Study Finds 4,100 Solar Jobs in Texas," Sierra Club press release, 2/14/2014; http://content.sierraclub.org/press-releases/2014/02/new-study-finds-4100-solar-jobs-texas.

[11]http://www.solarindustrymag.com/issues/Demo/FEAT_01_Municipal-Utilities-Ride-To-The-Rescue-Of-Beleaguered-Solar-Projects.html.

[12]http://newsroom.cpsenergy.com/blog/renewables/solar-renewables/solar-comparison/.

**Fig. 5.6** CPS Energy Blog

selling power for in New Mexico." Feinstein also said it is not common for projects to publicly release their rates, making comparisons that much more difficult.

A series of blog posts that ran under the banner "Powering Jobs in the New Energy Economy"[13] became a popular feature. The series profiled real San Antonians who had benefited from the new, high-paying jobs. Blog statistics also revealed that any time jobs were discussed on its Web pages, views to the site jumped (Fig. 5.6).

CPS Energy contracted with a local economics professor from San Antonio-based St. Mary's University to evaluate the overall economic benefit to the community. In quarterly reports, Dr. Steve Nivin tallied the economic impact to San Antonio of the NEE Initiative. Quantifying the economic development value for the community through a reputable third-party source and communicating that information repeatedly has been imperative to the initiative being recognized and appropriately valued by stakeholders.

## 5.7 Making It Happen

Identifying opportunities that fit the strategic objectives of the company, community, and environment was just the beginning. Finding the right partners and negotiating agreements was a major undertaking.

---

[13]http://newsroom.cpsenergy.com/?s=powering+jobs+in+the+.

CPS Energy typically issues an RFP to companies who will generate clean power. The companies will be responsible for constructing, operating, and maintaining all facets of the facility. CPS Energy will buyback the power from the companies with a Purchase Power Agreement (PPA). PPAs are typically 25–30 years. As a municipally owned utility, CPS Energy must adhere to strict purchasing requirements. Generally, an RFP must be created when there is more than one qualified vendor, such as solar manufacturers and system developers. Others, such as unique start-ups, may be eligible for a sole-source contract. Either requires diligence under normal circumstances, but under NEE program, the proposal review was complicated.

The NEE proposals and contracts required a more intricate and detailed process with some relevant exception policies added to corporate governance. They also required more time than anticipated. Neither the CPS Energy nor the respondents had experience negotiating major economic development in conjunction with contracts for services. The task required much thought, discussion, and open-mindedness on both sides to achieve success.

Noteworthy was the newness of the economic sector in San Antonio. Luring businesses and economic development is much easier with new companies and new business models that have shallower roots. Coaxing a large, established company with a mature business model to pack up shop and relocate to Texas would have been less likely than working with innovators. New companies are generally not deeply rooted, more flexible, and more willing to assume risk.

Yet CPS Energy was interested in companies with staying power. Finding the ideal combination of newness and financial strength was challenging. Start-ups are not typical business partners of municipal utilities in Texas. The companies selected had to have a progressive edge and a solid business plan for long-term stability.

## 5.8 Go Big or Go Home

CPS Energy asked much from its NEE partners: the manufacturing and installation of solar components, LED streetlights, home area networks, Smart Grid Technology, Advanced Metering Technology, relocation to San Antonio or establishment of a local office or plant, plus contributions to local STEM education initiatives.

Historically, as a typical utility, RFPs and other project purchases were made in increments as the annual budget allowed. With the hefty economic development requirements, the utility quickly learned that potential partners would demand a much bigger commitment. The PPA, which is for 25 years, was increased from 100 to 400 MW.

To meet those requirements, contracts had to have steps that benchmarked each progression in the agreement, almost like a series of small contracts bundled into one. This process required serious foresight into possible future states and conditions, with risk mitigations built into each milestone. Economic development milestones had to be met before the contractor would be allowed to progress to the

### San Antonio goes big on solar

San Antonio's municipal utility, CPS Energy, offers a variety of customer rebates to help achieve the goal of obtaining 20 percent of its electricity from renewable sources by 2020. The utility also has a 25-year agreement with San Antonio-based OCI Solar Power to purchase power from the 400-MW Alamo Solar Farm, which when completed in 2016 will be the largest in the nation. In addition to powering nearly 70,000 homes, the Alamo Solar Farm will generate $1 billion in construction investment, $700 million in annual economic output, and 800 permanent jobs. So far, OCI has completed the first three phases, with a combined capacity of 85 MW.

The Alamo 4 Solar Farm in Bracketville.

**Fig. 5.7** Pew Charitable Trusts[14]: CPS Energy is "leading the charge, establishing aggressive renewable energy goals"[15]

next set of work deliverables. Fortunately, with each new agreement, the process became easier.

Internal issues resulted from this new approach, as well. CPS Energy's legal department and consultants were challenged to develop contracts unlike any they had ever created before at the utility.

Forging relationships with local universities and colleges, economic development foundations, and local government staffers was essential to making these deals happen. Without a complete economic development package, there would not have been a compelling incentive for companies to locate here; elements included temporary office space, press coverage, location siting services, and, in some cases, financial incentives. These companies were more than contractors. They had become NEE partners (Fig. 5.7).

---

[14]http://www.pewtrusts.org/~/media/Assets/2014/11/Texas_11thHourBrief.pdf.

[15]"Texas Winds Generate Economic Growth; clean economy rising" Pew Charitable Trusts. November 17, 2014.

CPS Energy partnerships included a groundbreaking agreement with OCI Solar Power to supply 400 MW of solar power to San Antonio and the surrounding area and bring solar manufacturing to the region. The agreement mandated the creation of at least 800 jobs and a $100 million capital investment over five years. In 2014, the country's first N-type solar panel manufacturer, Mission Solar Energy, located its facility in San Antonio under incentives provided by CPS Energy. The highly efficient N-type solar panels lend themselves more readily to residential use where space is limited.

## 5.9 New Energy Economy Company Roster

From 2011 to 2014, CPS Energy partnered with seven clean energy companies to bring economic development to San Antonio. More companies followed as suppliers, totaling 11 thus far.

As a result, these companies have opened up manufacturing facilities, corporate headquarters, and local offices. Companies that have established corporate headquarters include Consert Inc., GreenStar, and OCI Solar Power along with suppliers for OCI Solar Power: Mission Solar, Sun Action Trackers, Mortenson, and KACO. Companies with local offices include Silver Springs and Landis + Gyr. Manufacturing facilities have also been established by Mission Solar, KACO, and GreenStar.

Details on how these and other partner companies have contributed to improving San Antonio follow.

**Consert**, a home area network (HAN) provider, currently employs close to 60 employees in San Antonio and is on their way to a projected 140 professional and technical jobs by 2017. The company's technology works to provide real-time energy use information and to allow aggregation for demand response. In 2010, CPS Energy and Consert launched a pilot program for 1,000 customers to start measuring electrical demand. Consert predicts the program will cut peak energy demand by 205 MWs if fully deployed. Currently, 20,000 installations have been completed with a potential 140,000 residential and commercial customers. Toshiba Corporation acquired Consert in 2013.

**GreenStar**, a global supplier of LED lighting, relocated its headquarters to San Antonio in 2011 and grew its staff to more than 40 in 2012. The company is working to install 25,000 LED streetlights across San Antonio over the next several years. The LED bulbs will save energy since each one uses about 60 % less energy than standard sodium lights. To further boost the economy, GreenStar contributed $10 from every LED bulb manufactured for CPS Energy to local education initiatives focusing on energy efficiency and renewable energy technology for a total of $250,000.

**OCI Solar Power** will provide CPS Energy customers 400 MWs of solar energy and its consortium anchor partner, Mission Solar Energy, just completed a $130 million manufacturing plant on the City's south side. The consortium will create more than 800 jobs with an average salary of $47,000. OCI Solar Power also moved its headquarters to San Antonio. The 400 MW are expected to be generated by seven sites around the state. As of November 2014, 90 MW have been constructed and are commercially available. When completed, the solar farms will provide enough electricity to power about 70,000 households, or about 10 % of the utility's customers. The OCI Solar Power consortium is expected to have an economic impact of roughly $1.3 billion per year in greater San Antonio, once all commitments are completed.

**Mission Solar Energy**,[16] a subsidiary of OCI Solar Power, became the first manufacturer of n-type solar PV cells and modules when it opened its doors in the USA in September of 2014. The highly efficient n-type solar modules are suitable for utility, commercial, and residential applications. The 240,000-square-foot facility on the former air force base now called BrooksCity Base can produce 2,000 modules per day and will employ 400 when fully operational.

**Sun Action Trackers**,[17] another OCI Solar Power partner, builds and assembles dual axis trackers, components in solar arrays that allow solar cells to gather the maximum amount of light for conversion to solar energy.

**KACO new energy**[18] makes solar PV inverters, the components that convert electrical current from DC to AC so it can be incorporated into the grid. Based in Neckarsulm, Germany, the company's 100,00-square-foot San Antonio plant now marks KACO's main North American manufacturing facility well positioned to deliver residential, commercial, and utility scale inverters throughout North and Central America. The headquarters moved to San Antonio from California in late 2013.

**Mortenson Construction** is the Engineering, Procurement, and Construction (EPC) company for the OCI Solar Power Consortium. The San Antonio office will be the southern US hub for all energy-related business including Mortenson's High Voltage Transmission Group, Wind Energy Group, Solar and Renewable Energy Group, and Oil and Gas Group.

**SunEdison**, one of the largest US solar energy service providers, has added 30 MWs of new solar power to CPS Energy's generation portfolio at three sites in the service territory. CPS Energy will buy the energy produced by the SunEdison solar farms at fixed rates for 25 years. SunEdison contributed $750,000 to various local education programs, including Alamo Area Academies, San Antonio and Somerset ISD Foundations, KIPP Aspire Academy, UTSA, St. Mary's University, and the University of the Incarnate Word as a result of this solar transaction.

**Summit Power Group** and CPS Energy negotiated a power purchase power agreement in September 2014 through the Texas Clean Energy Project for the first

---

[16]http://newsroom.cpsenergy.com/blog/made-san-antonio-solar-panels/.

[17]http://newsroom.cpsenergy.com/blog/made-san-antonio-solar-panels/.

[18]http://newsroom.cpsenergy.com/blog/kaco-inverters-san-antonio/.

US-based power plant that combines both Integrated Gasification Combined Cycle and 90 % carbon capture and storage technologies. The plant, to be built roughly 15 miles west of Odessa, TX, is expected to be online in 2019 and will provide 200 MWs of clean coal electricity to CPS Energy. The agreement calls for Summit Power Group to stage educational workshops and roundtables addressing clean energy issues in San Antonio to elevate understanding of this emerging field.

**Silver Spring Networks** is helping develop CPS Energy's Smart Grid Initiative, by upgrading infrastructure and building a two-way communication system that will increase reliability, give customers greater control over their energy use, reduce outage times, save money, better integrate renewable power into the grid, and improve the environment.

**Landis + Gyr** Technology, Inc. will manufacture more than 700,000 smart electric meters with Silver Spring Networks' technology to further the Smart Grid Initiative.

## 5.10 The Impact

- 842 Jobs
- $1.4 billion economic impact ($1.6 billion based on commitments)
- $4.2 million for education ($23.4 million committed)

As of October 2015, CPS Energy's NEE Initiative has resulted in more than 842 new, full-time jobs, $52 million in payroll, $200 million in capital investment, and $4.2 million in education contributions. Economic analysis conducted by Steve Nivin, Ph.D., chief economist of the SABÉR Institute, concluded that economic impacts expand to $1.4 billion with multiplier effects taken into account.

Once all commitments are attained, the economic impact with multipliers is estimated at over $1.6 billion per year and $23.4 million in educational contributions. Multipliers include spending on supplies, materials, and transportation, as well as the personal spending of employees on homes, cars, groceries, and living expenses.

In tandem studies released at the close of 2014 spotlighting the success of the CPS Energy's NEE Initiative, the utility was recognized by the Pew Charitable Trusts and Cogent Reports for leadership in renewable energy development and serving as a "customer champion." In The Pew Charitable Trusts' "Clean Economy Rising Texas" brief, CPS Energy was recognized as a "leading the charge, establishing aggressive renewable energy goals," in the Lone Star State.

CPS Energy serves as a national model of how utilities can simultaneously ensure their own economic well-being, while also supporting their community and the environment. While so many companies struggle to meet all three pillars of sustainability, CPS Energy is meeting social, economic, and environmental goals all under one unique initiative.

The New Energy Economy has come a long way since its start in 2011. Surveying a room full of city leadership, government officials, and staff at the opening of the Mission Solar Energy manufacturing plant in September of 2014, CPS Energy CEO Doyle N. Beneby expressed pride and delight, "Not just for CPS Energy, but for the promise this facility holds to produce clean energy, great jobs and educational investment for all of San Antonio for years to come."

# Chapter 6
# DTE Energy: Water Management

**Patricia Ireland**

**Abstract** Thermoelectric power plants withdraw a significant amount of water to cool the steam that drives electricity-producing turbines. The total amount of water withdrawn from and then returned to a water system (water withdrawal) and the amount of water withdrawn and consumed through evaporation, effectively removed from a water system (water consumption), are two measures of the impact power plants can have on a water system. On industry average, approximately 97 % of withdrawn water for cooling is returned to the water source, with a slightly increased temperature. Approximately 3 % of the withdrawn water evaporates during the power plant cooling process and is removed from the local water environment. DTE Energy's Greenwood Energy Center in Michigan uses a recirculating canal for cooling, operating with minimal water withdrawals. Factors that influenced the decision to use a recirculating canal cooling at Greenwood included: the anticipated growth of electricity demand in a region with limited surface water and the relative availability of land in the region. The plant is located in the Great Lakes basin, where water availability is not typically expected to be a challenge. However, due to the plant's location and municipal water source, water use and consumption was a design and operational focus at its inception. Subsequent expansions to the cooling system increased the surface area and improved energy efficiency. These improvements in turn helped to ensure compliance, optimize water use, and led to increased water conservation. In addition to sustainable water management, Greenwood works toward sustainability through its safety program, wildlife habitat protection, deer management program, and designation as a Clean Corporate Citizen by the Michigan Department of Environmental Quality.

P. Ireland (✉)
DTE Energy, Detroit, USA
e-mail: Irelandp@dteenergy.com

## 6.1 Water Matters

Water quality and water availability are essential and of increasing concern as we face regional scarcity, quality degradation, and climate-related challenges. The pressure is on all of society to thoughtfully manage our water use and protect water quality.

Thermoelectric power plants withdraw a significant amount of water to cool the steam that drives electricity-producing turbines. The total amount of water withdrawn from and then returned to a water system (water withdrawal) and the amount of water withdrawn and consumed through evaporation, effectively removed from a water system (water consumption), are two measures of the impact power plants can have on a water system. On industry average, approximately 97 % of withdrawn water for cooling is returned to the water source, with a slightly increased temperature. Approximately 3 % of the withdrawn water evaporates during the power plant cooling process and is removed from the local water environment.[1]

As part of the United States Geological Survey (USGS), water withdrawal and water consumption rates have been monitored every five years since 1950. Figure 6.1 shows relative water use and consumption rates for the USA in 2005.[2]

Thermal electric plants represent 39 % of the water withdrawals and 3 % of the water consumption in the USA. On average, 19 gallons of water are withdrawn and just under 0.6 gallons consumed to produce 1 kilowatt-hour (kWh) of electricity.[3] For the average US household, this translates into 207,252 gallons of water withdrawn and 6544 gallons of water consumed, annually.[4]

Thermal electric plants withdraw and consume water in varying quantities based largely on their cooling system design and water management practices (Table 6.1).

Water availability and cost are two primary considerations when selecting the water cooling system for a power plant. Other factors influencing the selection are cooling capacity requirements, land characteristics and availability, alternative water sources and stakeholder concerns. Among traditional cooling system technologies, recirculating tower systems generally have the lowest water use and the highest cost. As of 2010, approximately one-third of US thermoelectric generation came from plants using recirculating cooling towers. New cooling technologies (such as air cooling) may further reduce water use and have become viable options when siting new plants in water-stressed regions.

Innovations around the source water for cooling systems can also reduce power plant use of freshwater. Of the roughly 1290 thermoelectric plants included in a

---

[1] Diehl and Harris [1].
[2] United States Department of Agriculture, Economic Research Service [2].
[3] Maupin et al. [3].
[4] United States Energy Information Agency [4].

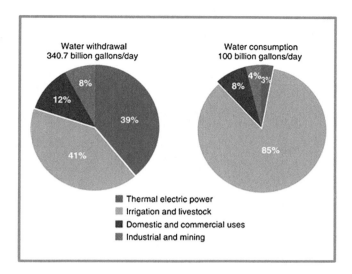

**Fig. 6.1** US water withdrawal and consumption, 2005

**Table 6.1** Water use and consumption rates by cooling system [5]

| Water cooling system (USGS 2010 data) | Number of plants in subset | Annual withdrawal (gallon/kWh) Average | Annual consumption (gallon/kWh) Average |
|---|---|---|---|
| Once-through lake/river/pond | 87 | 43.75 | 0.03 |
| Recirculating lake/pond | 10 | 1.57 | 1.57 |
| Recirculating tower | 304 | 0.85 | 0.60 |
| Greenwood (469,680 MWh) | 1 | 0.61 | 0.61 |

[a]Subset of plants with annual generation from 100,000 to 1,000,000 MWh's

2010 USGS survey, 86 plants reported using saline source water for cooling and 44 plants reported using public discharge water (typically wastewater).[5]

Installation of less water intensive cooling and generation technologies, improvements to existing cooling systems and innovations in sources for cooling water for both new and existing plants are responsible for an almost 20 % decline in thermal electric plant freshwater withdrawals from 2005 to 2010 (Table 6.2).[6]

---

[5]Diehl and Harris [5].
[6]Barber [7].

**Table 6.2** US thermal electric plant withdrawal and consumption rates, by cooling system [5]

| Water cooling system (freshwater) | Annual withdrawal (gallon/kWh) Average | Annual consumption (gallon/kWh) Average | Annual generation (GWh) | Number of plants |
|---|---|---|---|---|
| Once-through river | 41.12 | 0.027 | 740,753 | 172 |
| Once-through lake/pond | 38.00 | 0.025 | | 95 |
| Complex (combination of fuel types and cooling systems) | 16.61 | 2.57 | 781,815 | 138 |
| Recirculating lake/pond | 2.38 | 2.38[a] | 202,003 | 38 |
| Recirculating tower | 0.77 | 0.53 | 1,511,069 | 747 |

[a]Calculating consumption rates is an inexact science but the USGS considers the amount of water added to a recirculating cooling system to maintain water levels, the "makeup" water, as both the water withdrawal and the water consumed [6]

## 6.2 Greenwood Energy Center

DTE Energy's Greenwood Energy Center (Greenwood) (Fig. 6.2), located on almost 1800 acres in St. Clair County, Michigan, has a capacity of 785 MW, can operate on natural gas or fuel oil, and uses a recirculating canal for cooling. The plant began producing electricity in 1979 and currently employs approximately 60 full-time personnel. Factors that influenced the decision to use a recirculating canal cooling system at Greenwood include the anticipated growth of electricity demand in a region with limited surface water, the relative availability of land in the region, and the anticipated use of the plant primarily as a part of the capacity market. The plant is located in the Great Lakes basin, where water availability is

**Fig. 6.2** Greenwood Energy Center

not typically expected to be a challenge. However, due to the plant's location and municipal water source, water use and consumption was a design and operational focus at its inception resulting in effective water management and high performance from the start.

USGS estimates Greenwood's water withdrawal and consumption rate in 2010 as 0.61 gallon/kWh. This withdrawal rate is lower than industry averages for all cooling systems. A more relevant comparison, among plants within a similar range of annual generation, shows again that Greenwood has a lower average water withdrawal rate among all types of cooling systems. (Table 6.1).

Measuring and estimating methods for water use and consumption vary among agencies and industries. The USGS estimating methodology calculates that Greenwood has an annual withdrawal and consumption rate of 0.61 gallon/kWh. the lowest among plants with recirculating lakes and ponds among plants with similar annual generation. Operational data and municipal water records from 2008 to 2014 indicate that Greenwood's average annual water withdrawal and consumption rate is actually even lower at 0.47 gallon/kWh. This low rate can be attributed to Greenwood's recirculating cooling canal design and subsequent improvements, water use optimization, and water conservation.

**Cooling System Design**

The cooling system at Greenwood, as originally built in 1970, was a 1.7-mile-long recirculating cooling loop, 160 ft wide, holding approximately 100 million gallons of water. The canal was designed to provide cooling capacity with the assistance of 176 aerating spray modules (Figs. 6.3 and 6.4).

In the 1990s, a 25 million gallon water impoundment basin was added to provide an alternative source of water to replace evaporated water loss in the cooling canal. The basin was christened "James Bay" in honor of the Greenwood Plant Manager who spearheaded the project. The water stored within is captured from precipitation, seasonal high flows from the Jackson drain, and very occasionally, city water.

**Fig. 6.3** Greenwood Plant

**Fig. 6.4** Three of four spray modules are operational in this section of the cooling canal

Around this same time, spray modules were nearing end of life and becoming a maintenance challenge. With the reduced number of operating spray modules and a cooling canal unable to adequately handle high-temperature days, Greenwood was forced to lower (derate) generating capacity. The generation derates, particularly during times of high customer demand (hot summer days) impacted the plant's viability and became a pressing concern for the plant staff. At the time, the options considered to enhance the plant cooling potential were a cooling tower, more spray modules, or increased retention time and surface area. Bill Howard (Work Management Specialist at Greenwood) said, "The cooling tower option was very expensive and the number of new spray modules required would use too much energy, so we decided to let Mother Nature do the work."[7] The Mother Nature option, to expand the cooling canal, had few barriers, since the land was already owned, available, and well suited to this use. Howard and a team of Greenwood employees decided on the remedy, developed the expansion design, presented their business case, and secured just enough funding to move forward. Howard describes the approval process in 2000 as challenging: "It was a tough sell, but everybody agreed we had to do something and the solution we proposed seemed the best way forward. The fact we used all Detroit Edison [now DTE Energy] labor was an added bonus" (Fig. 6.5).

When the 5-month expansion project was complete, the canal capacity increased to 150 million gallons with double the surface area, improving the plant's cooling capability and largely resolving the derates caused by insufficient cooling. As part of the expansion project, a new 40 million gallon secondary impoundment basin was constructed in the center of the loop. The basin captures and stores water sourced from the canal (during seasonal high levels) and precipitation. This improved Greenwood's ability to manage water levels within the canal. As a consequence, water is infrequently discharged from the cooling

---

[7]Personal Communication, June 2015.

**Fig. 6.5** Bill Howard, Greenwood work management specialist

system, and only as a final option to maintain cooling loop water levels. The average annual discharge from 2009 to 2014 was 53 million gallons with zero discharge in both 2011 and 2012.

With rented equipment, the help of a company engineer, and two employees from another DTE plant, Greenwood personnel not only conceived of the plan, but also performed all the work to turn the plan into reality. Ten employees were recognized for their work on this project with DTE Energy's Alex Dow Award, a prestigious company award which recognizes outstanding achievements related to company and industry operations and humanitarian activities in the community.

Enhanced management of the cooling loop water level has helped Greenwood minimize water purchases from the municipality and reduce water costs. Additionally, the plant installed a riparian buffer along the canal, with the goal of improving water quality, increasing bank stability, and providing additional habitat for wetland wildlife (Figs. 6.6, 6.7 and 6.8).

**Fig. 6.6** Riparian bank along the 2000 expanded cooling canal

**Fig. 6.7** Turkey Vulture nesting along the cooling canal buffer zone

**Fig. 6.8** Greenwood wetlands provide waterfowl habitat

**Fig. 6.9** Construction activity for the second canal expansion

A second expansion of the cooling system was recently completed in 2015 (Figs. 6.9 and 6.10). The 2000 expansion proved that increased retention time and surface area in the cooling canal improve water cooling effectiveness and reduced plant derates. However, more could be done to optimize Greenwood's cooling

**Fig. 6.10** Newly carved wetland using GPS directed sculpting technology

system. As with the 2000 project, Greenwood employees developed the plan and business case. However, the second phase expansion was a much larger endevour so the construction part of the project was contracted (to a local, Michigan based company). The 2015 expansion doubled the current cooling canal surface area and ensures adequate cooling for the plant to operate at full capacity in protracted summer heat, improves energy efficiency by further reducing the need for water aeration in the canal, and further ensures compliance with the plant's water discharge permit. As part of this expansion, 18–20 new acres of wetland were created. Bill Howard, one of only two employees still at Greenwood from the 2000 expansion team, was thrilled to be a part of the second expansion project. "Until now, the highlight of my 39 year career with the company had been working on the 2000 canal expansion, and now I get to see the project through to completion." said Howard. "I had thought about retiring but didn't want to miss this," he added.

- **Capacity** In recent years, Greenwood has generated about 5 % of the electricity it is actually capable of generating. The 95 % of electricity that goes ungenerated is referred to as *capacity*. Capacity plays an important role in meeting the electricity needs of DTE Energy customers, especially on hot summer days when customer use of electricity increases dramatically (peaks). The 2015 cooling system expansion will increase the recirculating canal to ensure that Greenwood has adequate cooling water to safely generate at full capacity (785 MWh's) during high-demand/high-temperature days.

  The US energy grid system has ample capacity for the present. However, expected near-term changes in the national generation portfolio (increased renewable and natural gas generation and retiring coal plants) are expected to increase the relevancy the capacity market has on the cost and availability of energy.

- **Water-Energy Nexus** The water-energy nexus describes the interdependence of water and energy. It takes a tremendous amount of water to mine and extract fuels, generate energy, and cool thermoelectric plants. It also takes energy to extract, treat, deliver, and use water (estimated to be between 4 and 19 % of US energy production[8]) (Table 6.3).

---

[8]Copeland [8].

**Table 6.3** Water-energy nexus

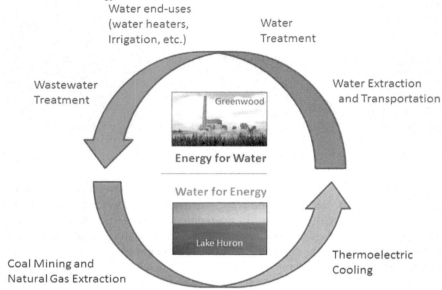

The interconnectedness of water and energy means that water use reductions and energy efficiency improvements have a compounding beneficial effect. The US EPA calls this the "Drops to Watts Connection"[9] and it is now part of the WaterSense Public Education effort. An example of this connection at Greenwood is the energy and water reductions expected as a result of the 2015 cooling system expansion. Greenwood currently uses approximately 142 floating spray modules to aerate the cooling canal and accelerate the cooling process (Fig. 6.11).

With the expansion complete, the need for the spray modules is reduced because of the doubling of the canal surface area. Greenwood is decommissioning half of the spray modules, saving approximately 1400 MWh/year of electricity and, in turn, over 600,000 gallons of cooling water annually that would have otherwise been used to generate that electricity.

The National Association of Regulated Utility Commissioners recently adopted a "Resolution Regarding the Water-Energy Nexus," recognizing this interdependency and urging state regulating bodies to enhance water and energy

---

[9]US Environmental Protection Agency [9].

**Fig. 6.11** Existing spray modules aerating the Greenwood cooling canal

efficiency efforts and explore compliance strategies that include the impacts of the water-energy nexus.[10]

- **Compliance** A consequence of using municipal city water as the primary source of makeup water for the recirculating cooling loop is a gradual buildup of fluoride. The incoming municipal water is fluoridated. Over the years of recirculation and evaporation in Greenwood's closed cooling canal, this level has become more concentrated. An April 2013 addition of a fluoride limit to the plant's water discharge permit of 2.7 mg/l has created a new compliance challenge for the plant. Greenwood is taking the following approach to comply: First, manage water levels so as to avoid permitted discharges. This involves maintaining the cooling loop level at the higher end of its operational range and temporarily halting inflows from James Bay (the makeup retention pond). Second, fill the new cooling loop expansion, as much as can be managed, with non-fluoridated surface water rather than city water. With this approach, Greenwood expects it can manage this compliance challenge.

**Water Use Optimization**

There are four sources for water inflow to Greenwood's recirculating canal (Fig. 6.12): the municipal water makeup valve, James Bay, the plant's secondary impoundment basin, and a gravitational feed from plant wastewater final polishing basin.

- Municipal water at the plant is supplied by the Detroit Water and Sewerage Department. The city has three water intakes throughout Michigan's Southeast region. The water delivered to Greenwood originates from a Lake Huron intake, is treated, fluoridated, and pumped to the plant via a high-volume pipeline installed specifically to serve the plant.
- James Bay is a naturally impermeable retention basin largely containing precipitation (estimated to be 119.5 million gallons/year). James Bay is also capable of receiving water from the municipal water supply in times of low water levels. During high water events, flow from a neighboring ditch (Jackson Drain) can

---

[10]National Association of Regulated Utility Commissioners [10].

**Fig. 6.12** Greenwood Plant

be directed to James Bay (estimated to be 10 million gallons/year). This estimated 10 million gallons/year is Greenwood's only direct withdrawal of water from the environment. As a natural, unlined basin, water only flows from James Bay into the cooling canal. The water in the canal is considered process water (because it is used for plant operations and potentially polluting materials) so cannot flow back to James Bay from the canal (Fig. 6.13).
- Greenwood's secondary impoundment basin is located downstream of the canal and operationally tied to the canal. Water can be pumped from the canal to the secondary impoundment basin and back again to the canal. This flexibility provides a final water elevation management tool and enables the plant to minimize water discharges while maintaining appropriate cooling canal water levels. Any discharge of water from the canal goes through the state permitted and monitored outfall pumped from the basin. The secondary impoundment also serves as a holding basin for sampling and analysis of water prior to discharge.
- Water and wastewater from plant operations is passively released to the cooling canal through a non-valved outlet pipe from a final polishing basin. The water flowing into the final polishing basin can come from any of the following

**Fig. 6.13** Aerial view of Greenwood energy center shows the cooling canal (before the start of 2015 construction), the secondary impoundment basin, and James Bay

sources: boiler blowdown, demineralizer regeneration waste, reverse osmosis wastes, oily wastewater, soot and ash sluice water, non-chemical metal cleaning waste, chemical metal cleaning waste, and sanitary waste. Virtually all water used by the plant, from the restrooms to the process water, finds its way the plant cooling canal. The flowchart lays out the main water sources within the plant and the path to the cooling canal (Table 6.4).

**Water Conservation**

Greenwood work procedures make clear the priorities in regard to city water use. In order, they are potable water, the turbine lube oil temperature, the condensate storage tank, the cooling canal, and James Bay.

In 2009, a project team researched canal water elevation and water bill history and the physics and assumptions built into the canal design, including its circulation pumps and cooling sprays. The team identified all the water loads in the plant and worked with the Detroit Water and Sewerage Department to better understand their data collection and billing factors. With this information in hand, the team focused on reducing the 12-month, peak-day intake of municipal water in ways protective of plant generation capacity and the five water priorities. As a result of this effort, the fixed portion of the water bill (based on projected peak-day water needs) was reduced by 46 %.

Greenwood employees convened a Water Council and established water use procedures that are now used to manage the plant's water use strategy. Digital

**Table 6.4** Greenwood Energy Center primary water flow

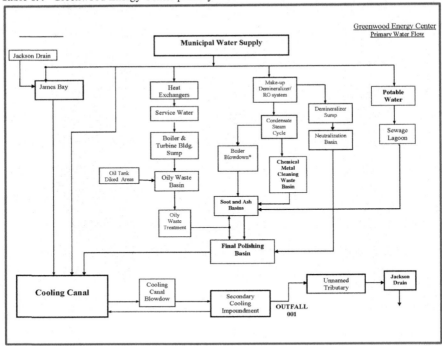

controls for the entire plant were installed in 2010. The cooling canal levels are now continuously monitored and trended using these digital controls. The real-time ability to monitor and predict canal levels enables the plant to fine-tune canal water management and is a primary tool in the plant's strategy of limiting city water use and cooling water discharges. Other water use load points within the plant are also now more easily monitored with the new controls.

In spite of water reuse and conservation practices built into the structural and nonstructural operations of the plant, water use reduction continues to be a priority of the plant. From 2008 to 2014, Greenwood paid an average of $2.17/MWh for municipal water. The average retail price of power in Michigan is $109.8/MWh.[11] The cost of water presents a very real driver to the plant to continue to effectively manage and minimize water use.

### Plant Sustainability Culture

In addition to a water management program designed and operated to minimize water use and preserve this important natural resource, Greenwood embraces a culture of sustainability across all areas of operation (Fig. 6.14). Inherent in the water management effort is a focus on energy reliability, energy affordability, and

---

[11] US Energy Information Agency.

**Fig. 6.14** Greenwood is home to a population of Bald eagles

habitat protection, all material issues for the electric utility industry. Our culture of sustainability is also evident in our focus on employee safety and health, community outreach, and stakeholder engagement. Some examples of this culture include:

- Safety

In 2013, Greenwood celebrated a 30-year milestone since experiencing a lost-time injury. This streak of safe work continues today. The small close-knit workforce at the plant gives Greenwood an inherent safety advantage. Plant Manager at the time, Mark Vander Heuvel said, "The number one comment I hear when it comes to safety is that we've got each other's back. The motto, '200 % accountable for safety,' is not new to Greenwood."[12] Greenwood achieved an overall score of 93 % in a 2013 National Safety Council survey, performing especially well on management communications, employees following safety procedures, management listening to employee safety suggestions, and the importance of safety versus production. Areas for continued improvement are effectiveness of award programs and employee participation in incident investigations.

- Wildlife Habitat Protection

Greenwood achieved Wildlife Habitat Certification from the Wildlife Habitat Council (WHC) in 2004. Approximately 1780 acres of the facility is managed for habitat and species protection. The land is composed of grasslands intermixed with stands of conifers and deciduous trees, wetland areas, and a number of water drainage and storage features. Facility employees have created a pollinator garden, food and cover plots for wildlife, and nature trails, which provide access to the

---

[12]Personal Communication, June 2013.

site's diverse habitats. Employees conduct educational programs geared at helping children understand their local ecology and the natural world and host the Audubon Society Annual Bird Count (Figs. 6.15, 6.16, and 6.17).

- Deer Management Program

Greenwood has an employee directed deer population management agreement with the Michigan Department of Natural Resources. As a part of this agreement, the plant is issued a number of deer hunting permits (32 in 2014). Greenwood employees may obtain one of the permits by participating in an internal lottery but

**Fig. 6.15** Wildlife habitat protection

**Fig. 6.16** Greenwood employees install bat boxes on sidewall of the plant

**Fig. 6.17** Greenwood employees monitor nesting boxes

**Fig. 6.18** Deer management program

must first demonstrate bow proficiency and safety practices in a test administered by fellow employees. All deer meat obtained through the bow hunting is donated to a food bank in Port Huron, Michigan (Fig. 6.18).

- Environment and Community

The Michigan Department of Environmental Quality designated Greenwood a Clean Corporate Citizen in 2009. The Director of the Michigan DEQ cited, "the development and implementation of an environmental management system, the practice of pollution prevention, and a strong environmental compliance record" as criteria for awarding the designation. The plant has been third-party certified to the ISO 14001 standard since 2004. Greenwood also participates in many local community programs, such as Keep America beautiful adopt-a-highway, United Way of St. Clair County, the Yale Food bank, and Salvation Army (Figs. 6.19 and 6.20).

**Fig. 6.19** Environment and community

**Fig. 6.20** Greenwood Employees engage with the local community

## 6.3 Conclusion

Greenwood's innovative approach to water management and culture of sustainability places it among the top performers in the industry. Through optimizing and improving the water cooling system design, implementing water use (and reuse) processes, and employing water conservation efforts, Greenwood's water withdrawal rate (and consumption rate[13]) is 0.47 gallon/kWh. This is better than the average rate within each cooling technology category tracked by the USGS. Greenwood's culture of sustainability, exemplified by a focus on reducing operational costs; strong safety record; certified ISO 14001 environmental management system; wildlife habitat protection program; and community engagement, has driven Greenwood's performance and kept the plant on good terms with neighbors and other stakeholders and enabled the plant to remain economically viable. Greenwood is clearly a part of DTE Energy's portfolio now and into the future.

Greenwood's low water withdrawal and consumption rate can be attributed to the large tract of land purchased at the time the plant was sited, favorable geology for cooling canal containment, a pervasive culture of cost (and water) minimization, and a positive relationship with neighbors and local government. Plants interested in implementing similar water efficiencies will need access to enough land to accommodate adequate cooling canal retention time, a strong driver for optimization of costs (and all sources of water), and a culture of strong employee engagement. Any plant possessing these attributes can replicate Greenwood's success at sustainable water management.

## References

1. Diehl TH, Harris MA (2014) Withdrawal and consumption of water by thermoelectric power plants in the United States 2010. Scientific investigations report 2014-5184, 13. US Geological Survey, Reston, Virginia
2. United States Department of Agriculture, Economic Research Service (2015) Agriculture is a major user of ground and surface water in the United States. http://www.ers.usda.gov/data-products/chart-gallery/detail.aspx?chartId=40024&ref=collection&embed=True. Accessed June 2015
3. Maupin MA et al (2014) Estimated use of water in the United States in 2010. Thermoelectric Power, 40. US Geological Survey, Reston, Virginia Circular 1405
4. United States Energy Information Agency (2013) Frequently asked questions web site. Household electricity use. http://www.eia.gov/tools/faqs/faq.cfm?id=97&t=3. Accessed June 2015
5. Diehl TH, Harris MA (2010) Withdrawal and consumption of water by thermoelectric power plants in the United States. Scientific investigations report 2014-5184, Appendix 1. US Geological Survey, Reston, Virginia

---

[13]Diehl and Harris [6].

6. Diehl TH, Harris MA (2010) Withdrawal and consumption of water by thermoelectric power plants in the United States, 2010. Scientific investigations report 2014-5184, Recirculating pond withdrawals, 12. US Geological Survey, Reston, Virginia
7. Barber NL (2014) Summary of estimated water use in the United States in 2010. Science for a changing world, United States Geological Survey. ISSN 2327–6916
8. Copeland C (2014) Energy-water nexus: the water sector's energy use. Congressional research service, 7-5700 R43200. 3 Jan 2014
9. US Environmental Protection Agency. Make the drops to watts connection. Publication EPA-832-K-08-001
10. National Association of Regulated Utility Commissioners (2014) Resolution regarding the water-energy nexus. Adopted by the NARUC Committee of the Whole, 19 Nov 2014

# Chapter 7
# Duke Energy: Engaging Employees in Sustainability

**Michelle Abbott**

**Abstract** Duke Energy serves 7.3 million retail electric customers in six states in the Southeast and Midwest regions of the United States. To be successful, we need our 28,000 employees to be fully engaged in delivering electric services in a sustainable way—affordable, reliable and clean. To help accomplish this, Duke Energy initiated an effort in 2008 to develop an approach and set of internal programs to engage employees in sustainability. Early results were positive and our goal of driving tangible business benefits was being achieved. Our merger with Progress Energy in mid-2012 provided an opportune time to reevaluate our approach in light of the company's much larger footprint. We found that the original design of the approach was still sound, but did make a number of enhancements to the program. As we continue to face unprecedented change in our industry, the company will continue to adapt its approach to engaging employees in sustainability, leveraging what we have already learned about unlocking employee ideas and innovation.

## 7.1 Introduction

For a company to be truly sustainable, sustainability has to be embedded throughout the organization. It cannot be the sole responsibility of a corporate department. Just like safety, sustainability has to be everyone's responsibility.

Duke Energy recognized this back in 2008 when it initiated an effort to develop an approach and set of internal programs to engage employees. The goal was to

---

M. Abbott (✉)
Duke Energy, Charlotte NC, USA
e-mail: michelle.abbott@duke-energy.com

build a sustainability culture at Duke Energy that engages employees, unlocks innovation, and delivers tangible business value to the company. After the first wave of implementation, the results were positive and the program was starting to gain traction.

In mid-2012, Duke Energy merged with Progress Energy to become the largest investor-owned electric utility in the USA. Instead of 18,000 employees, the company now had 28,000. The merger provided an opportune time to relook at the engagement approach in light of the company's much larger footprint.

This chapter will describe how Duke Energy initially developed its employee engagement programs, the refinements we made to it along the way, the tools we have created, and the lessons we learned. Like many other companies, we still have a ways to go, but we are happy to share the road we have travelled thus far.

## 7.2 Laying the Foundation

Similar to many other electric utilities, Duke Energy's commitment to sustainability is a natural extension of its long history of environmental stewardship and community involvement. As an electric utility, we have a large impact on the environment and our success is linked to the health and prosperity of the communities we serve.

In the late 1990s and early 2000s, early sustainability efforts were spearheaded by the Corporate Environment, Health, and Safety department. In mid-2006, just after the Duke Energy merger with Cinergy, a small, separate sustainability department was created. Our Chairman, President and Chief Executive Officer at the time, James E. Rogers, was an early adopter of sustainability and provided visible support for the newly formed department.

Given Cinergy and Duke Energy had both participated in the annual Dow Jones Sustainability Indices (DJSI) assessment several times before the merger, the new department adopted the DJSI definition of sustainability to guide its efforts:

> "Corporate sustainability is a business approach that creates long-term shareholder value by embracing opportunities and managing risks deriving from economic, environmental and social developments."[1]

While the DJSI definition served as a good starting point for the department's efforts, it was too cumbersome to be easily digested by people outside the department. It was certainly not something one could easily explain during a brief elevator ride and therefore did not pass "the elevator test." Given this, we developed a much simpler definition that was easier to remember and grasp:

> "At Duke Energy, sustainability is doing business in a way that is good for people, the planet and profits."

---

[1]"Corporate Sustainability," Dow Jones Sustainability Indices, accessed June 19, 2015, http://www.sustainability-indices.com/sustainability-assessment/corporate-sustainability.jsp.

This new definition—which Duke Energy still uses today—is a slight variation of the "triple-bottom-line" concept that was first coined by John Elkington, a well-known authority on corporate sustainability, back in 1994.[2]

In addition to getting a definition for sustainability in place that resonated for Duke Energy, the newly created department focused on developing the company's first corporate sustainability plan and goals, the first sustainability report, and the first application to the DJSI as a merged company.

We partnered with key internal departments to ensure initiatives were underway to support the newly created sustainability plan and goals (there were 30 goals initially). Developing the first sustainability report and the first DJSI application as a merged company also helped get many conversations going on internally.

In some areas of the company, there was genuine excitement. In other areas, there was healthy skepticism that "this too shall pass," just like some of the other corporate programs that had come before it. Regardless, the wheels were turning and our sustainability journey had started.

## 7.3 Building an Employee Engagement Program

By 2008, the sustainability department was ready to build on the foundation that had been laid during the first couple of years. The inspiration to build an employee engagement program came when our CEO, James E. Rogers, read about Walmart's new employee engagement program. If Walmart could successfully engage its employees on sustainability given its many locations and huge workforce, surely Duke Energy could as well. To find out, we began conversations with the consulting agency that worked with Walmart on their employee engagement program, Act Now, which became Saatchi and Saatchi S.

The first step was to partner with Saatchi and Saatchi S to develop a detailed scope of work and project plan. The project plan included a current state assessment, followed by a detailed design of the new program. Once the project plan was in place, we signed a contract with Saatchi and Saatchi S to begin the development of the new employee engagement program.

Based on what Duke Energy had already learned from its efforts to build a safety culture, we knew we wanted our program to include an emphasis on sustainability on and off the job. To help guide the development of Duke Energy's program, we established an internal steering team comprised of mid-level executives from across the company. As plans and concepts were developed, we shared them with the internal steering team to get their input and feedback.

An important part of the current state assessment was when representatives of Saatchi and Saatchi S interviewed 32 Duke Energy senior executives to get their frank feedback on sustainability. Several executives admitted they were not used

---

[2] "Triple Bottom Line," *The Economist*, November 17, 2009, accessed June 19, 2015, http://www.economist.com/node/14301663.

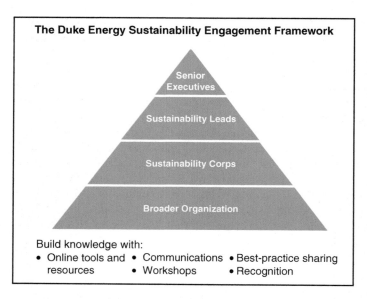

**Fig. 7.1** The Duke Energy Sustainability Engagement Framework

to the word "sustainability," but they did largely support the concept given it built upon the company's traditional focus on communities and the environment. The different cultures within Duke Energy (e.g., a power plant versus the corporate office) were mentioned as a key challenge to overcome. In addition, we were cautioned that "program fatigue" and skepticism would likely be encountered.

To round out the current state assessment, representatives of Saatchi and Saatchi S made several field visits to see our operations in action and reviewed many existing company communications, employee survey results, policies, and practices. This helped generate understanding of the existing practices and culture at Duke Energy. Once the current state assessment was complete, we proceeded to jointly design what ended up being a multi-tiered approach to educating and engaging employees on sustainability.

Figure 7.1 illustrates the multi-tiered approach that was developed. It includes engagement strategies for each layer of the triangle, supported by a foundation of tools and other resources.

## 7.4 Starting at the Top

Duke Energy's Executive Leadership Team includes the direct reports to the company's Chief Executive Officer and their direct reports (approximately 60 individuals in 2009). This group of senior executives met quarterly, providing an opportune time to engage the company's top leaders. Typically, the agenda for the daylong meeting included several different topics, but in the first quarter of 2009, the entire meeting was devoted to sustainability. The goal was to further the executives' understanding of sustainability, plus gain their support and vision. We also wanted to get their feedback on Duke Energy's overall approach, including the newly designed approach for employee engagement.

The meeting began with a panel of prominent, external speakers:

- Jonathan Lash, President, World Resources Institute
- Neil Hawkins, Vice President, Sustainability, Dow Chemical Co.
- Andrew Howard, Executive Director, Global Investment Research, Goldman Sachs International
- Steven Fludder, Vice President, Ecomagination, GE
  (Note: These were their titles and organizations at that time.)

The external speakers provided real-life examples of how a focus on sustainability had driven business value at their organizations. Hearing about other companies' successes, especially the significant cost savings and revenue increases that had been realized, helped bolster credibility and support with Duke Energy leadership. After a robust question and answer session, the panelists departed and the focus was turned inward.

To facilitate frank discussion, we divided the executives into small groups. Each group included seven-to-eight executives, which was small enough to fit at one table. Their first assignment was to brainstorm the top challenges and opportunities facing Duke Energy and how sustainability thinking could help address them. Each small group appointed a spokesperson and reported their ideas to the full group. Energy efficiency, emerging technology, and the need for new regulatory models were just a few of the topics that were discussed.

Next, the small groups brainstormed improvements needed to our sustainability plan and goals. Again, each small group reported back its findings to the full group. The need to substantially reduce the number of sustainability goals (they had grown from the initial 30 to 40 by this time) and make them more quantitative in nature was a resounding message.

The newly designed approach for engaging employees was then shared and discussed with the executives (see Fig. 7.1). The executives liked that it was both top-down and bottom-up. They also liked that the program was not highly prescriptive and flexible enough that it would work in a power plant as well as a corporate office setting. The effort that had gone into the current state assessment clearly paid off. The executives were largely supportive of the proposed approach.

Our CEO closed the Executive Leadership Team meeting by challenging the executives to keep the support and momentum for sustainability going. Key executives left the meeting with an action item to formally name at least one Sustainability Lead for their department. This was the formal launch of Duke Energy's employee engagement program.

## 7.5 Level Two: The Sustainability Leads

Sustainability Leads, or "S Leads" for short, are the second tier in Duke Energy's approach to employee engagement. Their mission is to integrate sustainability into their departments to further Duke Energy's progress as a best-in-class sustainable company. Given there is no one size fits all solution, it is key for them to integrate sustainability in a way that makes sense for their departments (Fig. 7.2).

S Leads are selected by top management based on the following criteria:

- High-potential employee who is capable of delivering results and interested in demonstrating leadership
- Trusted and respected by subordinates, peers, and department management
- Has the ability to think beyond his or her work group
- Passionate about making a difference and doing business in a way that is good for people, the planet, and profits

The initial charge of the S Leads was to identify and work on three key opportunities in their departments, plus help recruit Sustainability Corps members (Level Three).

We believed at the outset that every major department—corporate and operational—needed to have an S Lead. This resulted in about 40 S Leads being named, which was a fairly large group to manage. To equip and train the S Leads, we partnered with Saatchi and Saatchi S to design and facilitate a six-hour workshop.

After the initial set of S Leads were trained, we touched base with them intermittently to see how their efforts were coming along. Since one staff member was managing the program, it was difficult to spend one-on-one time with every one of them on a frequent basis. However, a quarterly conference call with all of the S Leads was held to share status updates and best practices. S Leads with good initiatives underway were asked to describe their efforts and lessons learned.

"Every minute we operate our nuclear plants contributes immensely to sustainability. Our fleet safely provides affordable, carbon-free electricity 24/7. However, we know we can do even more in the area of sustainability, and we strive to do so."

Steve Nesbit | Director, Nuclear Policy and Support, Nuclear Corporate Office S Lead

**Fig. 7.2** A Duke Energy Sustainability Lead

## 7.6 Level Three: The Sustainability Corps

The third tier in Duke Energy's approach to employee engagement is the Sustainability Corps. The Sustainability Corps, or "S Corps" for short, is a grass-roots network of Duke Energy employees. Its mission is to further our progress as a best-in-class, sustainable company. The name was inspired by the Peace Corps, the famous volunteer US government program that also has a noble mission.

Any employee is eligible to join the S Corps as long as they have their manager's support. The program provides employees who have a passion for sustainability a way to make a difference. This can be especially appealing to recent college graduates who tend to have a strong interest in sustainability.

An ideal candidate for the S Corps is:

- Excited to learn more about sustainability and its relevance to the company
- Interested in helping Duke Energy do business in a way that is good for the "three Ps"—people, the planet, and profits
- Willing to develop new skills and accept leadership responsibilities, including leading work-related projects

Like the S Leads, Corp members attend a workshop to educate and equip themselves on sustainability. They are then asked to:

- Adopt a Personal Sustainability Practice (PSP), which is a simple act the participant commits to do on a regular basis in their personal life that is good for sustainability
- Lead at least one project to help integrate sustainability into their work area

Once the workshop is completed, the employee becomes a member of the S Corps (Fig. 7.3). Afterward, they attend follow-up teleconferences with their class and the workshop facilitator. These calls initially occurred around 2, 5, and 12 weeks after the workshop. The purpose is for new S Corps members to provide updates on how they are doing on their PSPs and work-related projects.

"Being a S Corps member has taught me that sustainability begins with asking the right questions. Continually challenging the status quo uncovers hidden costs and benefits."

Lisa Smith | Program Manager, Inventory Optimization, Supply Chain, S Corps Member

**Fig. 7.3** A Duke Energy Sustainability Corps Member

As was mentioned earlier, we relied on the S Leads to help recruit S Corps members. This was very instrumental in getting the S Corps program going initially and provided a boost to the company-wide communications that also went out about the new program.

Our Human Resources department also partnered with us to get the word out about the S Corps program. They added the S Leads and Corps workshops to Duke Energy's training and development catalog to increase awareness of the workshops. Human Resources also set up the process to ensure employees get credit on their training transcripts for taking the classes.

Information about the S Corps program was also included on the sustainability page of the company's intranet site. Employees who are interested in joining can also register there for an upcoming workshop.

## 7.7 Home Grown Workshops and Tools

We partnered with Saatchi and Saatchi S to develop the S Leads and Corps workshops. The goal of the workshops was to educate participants on what Duke Energy is doing to become a more sustainable company, plus equip them with practical tools that they can apply within their departments.

Given they tend to be higher in the organization and their focus is the entire department, we expect the S Leads to be a bit more strategic with their improvement ideas than the S Corps members. At the same time, innovation and good ideas exist at every level of the organization. Given this, the tools we created for the workshops were intended to be applicable to both the S Leads and the Corps members.

### 7.7.1 The Energy Tool

The purpose of the first tool that we developed for the workshops, the Energy Tool, is to help participants select a PSP (Fig. 7.4).

Typical examples of PSPs include: avoiding landfill waste, printing fewer documents, volunteering in the community, composting, using a reusable water bottle, taking alternative transportation, exercising more, and planting a garden. One employee even went so far as to preserve and protect family property that he had inherited so its natural beauty could be enjoyed by future generations.

The Energy Tool is a simple checklist to help participants think through a good PSP for them.

## Duke Energy's PSP Energy Tool

When choosing a Personal Sustainability Practice (PSP), remember that it should have "ENERGY." Your PSP should be:

- (E) Environmentally-friendly – something that helps sustain the planet
- (N) New – a change in your normal routine
- (E) Engaging – something you can share with others
- (R) Repeatable – something you can do every day or a few times a month
- (G) Good – for you, your family, your finances or your community
- (Y) Yours – something that relates to your life and your interests

**Fig. 7.4** Duke Energy's PSP Energy Tool

### 7.7.2 The Sustainability Filter

The second tool we developed for the S Leads and Corps members workshops was the Sustainability Filter (Fig. 7.5). It enables participants to view decisions and actions through the lens of sustainability.

The Sustainability Filter includes four key concepts that are fundamental to sustainability:

- connection,
- efficiency,
- balance, and
- grandchildren.

Each of the four key concepts is defined briefly in the tool. In addition, a series of 3–4 questions is provided for each of the four key concepts to help enable more sustainable decision making.

During the workshop, participants are divided into small teams and given the exercise of applying the Sustainability Filter to a past complex decision that Duke Energy has had to make. Examples include:

- how to offer our customers more energy efficiency options without harming profits,
- whether or not the company should repair or close the Crystal River Nuclear Plant, and
- how to address stakeholder concerns about our purchasing of coal mined through mountaintop removal. (This is a controversial surface-mining technique that is accomplished by removing the tops of mountains in order to reach coal seams.)

| Duke Energy's Sustainability Filter |
|---|
| Connection: Understanding the big picture and the interrelationships between issues<br>• Have we considered the financial, environmental and social impacts of this action/decision?<br>• Have we taken potential changes in the external environment, such as new regulations, into account?<br>• Have we considered this action/decision in light of our key stakeholders' expectations and priorities? Have we looked for the connections between issues?<br>• Have we examined it from a life cycle/value chain perspective? |
| Efficiency: Using resources as efficiently as possible to save money and respect our planet's limits<br>• Does this action/decision help us reduce our use of resources — materials, energy, water, etc.? What about our suppliers? Customers?<br>• Does it help us improve our performance on the 3Rs of solid waste (reduce, reuse, recycle)? What about our suppliers? Customers?<br>• Does this action/decision provide us an opportunity to profit from what we might otherwise throw away? |
| Balance: Developing solutions that effectively address competing interests<br>• Does this action/decision balance our stakeholders' competing priorities?<br>• Does this action/decision balance "people, planet and profits?" Can we develop a win-win-win solution?<br>• Does it balance short-term and long-term needs?<br>• Have we evaluated purchases and performance of suppliers against these same questions? |
| Grandchildren: Anticipating how future generations will view the actions we take (or don't take) today<br>• Have we looked at this action/decision through the eyes of future generations?<br>• Will it stand the test of time?<br>• Will this action/decision contribute to long-term shareholder value?<br>• Will it benefit, or at least not harm, society and the environment? |

**Fig. 7.5** Duke Energy's Sustainability Filter

This exercise helps participants better understand the complicated, connected world that we live in and how challenging it can sometimes be to address sustainability issues. Participants also learn the importance of asking key questions upfront and considering the perspectives of Duke Energy's diverse stakeholders while making decisions.

The Sustainability Filter is a good tool for the S Leads and Corps members to use as they evaluate decisions and actions, plus share with others in their departments. The tool not only helps with complicated matters, but can also help with everyday decisions and actions.

## 7.7.3 The Opportunity Finder

The third tool we developed for the S Leads and Corps workshops was the Opportunity Finder (Fig. 7.6). Similar to the Sustainability Filter, participants can use the Opportunity Finder themselves, plus share with others in their departments.

| Duke Energy's Opportunity Finder ||
|---|---|
| **Instructions:** <br> • Gather a team and use this template to brainstorm key sustainability opportunities in your department. Record potential opportunities in the appropriate cells. <br> • After identifying the opportunities, evaluate them in terms of positive impact and probability of success. <br> • Determine the top opportunity and develop an action to address it. ||
| *Brainstorm potential opportunities to:* ||
| **People (Social)** ||
| Improve workforce safety, health or wellness | |
| Improve reliability and/or the customer experience | |
| Foster workforce diversity and inclusion | |
| Engage or develop employees | |
| Encourage employee volunteerism and giving | |
| Support our communities/foster economic development | |
| Partner with suppliers on sustainability | |
| **Planet (Environment)** ||
| Reduce energy/fuel use and greenhouse gas emissions | |
| Reduce other air emissions | |
| Reduce water use and/or improve water quality | |
| Reduce waste, materials and natural resource use | |
| Increase renewable energy use | |
| Protect habitat and biodiversity such as fish and wildlife | |
| Reduce use of chemicals | |
| **Profit (Economic)** ||
| Reduce or eliminate costs (energy, water, fuel, paper, chemicals, materials, equipment, vehicles, disposal costs, labor) | |
| Grow revenues (expand the use of existing products or develop new ones; make money from what we currently throw away) | |

**Fig. 7.6** Duke Energy's Opportunity Finder

The Opportunity Finder is intended to help participants identify sustainability opportunities within their departments. It is organized by the "three Ps"—people, the planet, and profits—and includes a listing of impact areas within each category to help facilitate brainstorming of possible sustainability opportunities.

Once the opportunities are brainstormed, they are prioritized based on positive impact and probability of success. Action plans are then developed for the opportunities with the most potential.

The Opportunity Finder is the key tool participants use to identify and select good work-related projects. Once these projects are underway, the Sustainability Filter can be used as a "check" to help ensure the actions being considered will be sustainable long term.

### *7.7.4 Online Tracking Tools*

To facilitate best practice sharing across the company, we worked with our Information Technology department to develop two online tracking tools. The first one, the PSP Tracker, simply includes fields for the submitter's name, his/her PSP, and the category it falls into (e.g., waste, water, emissions/energy reduction, process improvement, community). If an S Corps member is struggling to come up with an idea of their own, he/she can get inspiration from others' PSPs by looking at the tracking tool. The tool also enables the sustainability department to see what topics are of most interest to S Corps members.

The second online tracking tool, the Project Tracker, is more robust and includes fields for project description; start/stop dates; project type; and how the project will benefit people, the planet, and profits (including expenses and savings). Tips for planning and prioritizing the project and sample calculations are also provided. Like the PSP Tracker, the Project Tracker provides a means to collect sustainability information and share best practices.

## 7.8 The Base of the Triangle: The Broader Organization

The fourth and final tier in our approach to employee engagement is the broader organization, which includes the rest of Duke Energy. To educate and engage the broader organization, we have primarily relied on the company's intranet site and other communications vehicles (e.g., weekly newsletters, periodic manager updates) to get the word out about sustainability.

The sustainability page on Duke Energy's intranet site includes a message from our CEO and a large number of resources—including the tools described earlier—and helpful links to external materials. In addition, as S Leads and Corps members complete work-related projects, we produce periodic articles or short videos to highlight those that are especially successful (Fig. 7.7).

"Sustainability means different things to different people. To me, it comes down to enduring business success and responsible stewardship. We'll do this by understanding stakeholders' needs, running our business with excellence and adapting to an ever-changing industry."

Lynn Good | Chairman, President and Chief Executive Officer

**Fig. 7.7** Message from the CEO

Another way we engage the broader organization is a popular sustainability quiz we launch internally each year just after the sustainability report is released. The quiz includes 10 or fewer multiple choice questions that focus on the sustainability report's key messages. The final question asks employees for their ideas and feedback on how Duke Energy can be more sustainable.

The quiz runs for four weeks and each week an employee who has completed the quiz is randomly selected. The winners receive a $100 gift card for themselves and a $100 gift card for the charity of their choice. Once the quiz is complete, we analyze the comments and suggestions for follow-up actions. Typically, about 7–9 % of Duke Energy employees participate in the annual sustainability quiz.

In 2011, we developed a ten-minute video entitled "Change Begins With Us" providing context on the sustainability challenges facing Duke Energy and our industry. We featured several S Leads and Corps members and had them describe successful efforts they had undertaken to help the company be more sustainable. The video also included vignettes from members of upper management to demonstrate their support. The video closed with information on the next steps employees could take to become more involved in sustainability, including becoming a S Corps member.

## 7.9 Gearing up for the Progress Energy Merger

In mid-2012, Duke Energy merged with Progress Energy to become the largest investor-owned electric utility in the USA. While improvements and refinements had been made to the sustainability employee engagement program over the years since it was launched in 2009, the merger provided an opportune time to relook at the approach in light of the company's much larger employee base and footprint.

## 7.10 S Lead Program Enhancements

As was mentioned earlier, the S Leads' mission is to integrate sustainability into their departments to further Duke Energy's progress as a best-in-class, sustainable company. A review of 2009–2012 results revealed that the biggest benefits (e.g., cost savings, avoided air emissions, increased recycling, and reduced chemical use) were being driven by the S Leads in the operational units. This made sense in hindsight because our power plants and power delivery field operations tend to have the highest costs and environmental impacts.

We also realized that the sustainability department already had other well-established pathways for partnering with corporate departments, namely via responding to the annual Dow Jones Sustainability Index (DJSI) assessment. Specifically, we rely heavily on many corporate departments to help us complete the annual DJSI questionnaire, plus identify possible improvements to our practices and performance. In addition, a review of the S Corps roster indicated the majority of the members worked in corporate departments.

Given these experiences and insights, we decided to focus the S Leads program solely on the operational units. After the merger with Progress Energy was completed, we asked executives over the operational units to confirm or name a new S Lead for their department (Table 7.1).

This brought the total number of S Leads down from 40 to 24, which is a much more manageable size. We also assigned a "Coach" from the sustainability department to each of the 24 Sustainability Leads to facilitate more one-on-one collaboration. While one member of the sustainability department provides overall

**Table 7.1** Duke Energy Sustainability Leads Per Operational Unit

| Operational Unit | Number of Sustainability Leads |
|---|---|
| Nuclear generation | Seven (one for corporate and one for each of six nuclear stations) |
| Fossil generation | Four (one for each of four geographic regions) |
| Renewables generation[a] | Two (one for corporate and one for operations) |
| International generation[a] | One |
| Transmission | One |
| Distribution | Three (one for each of three geographic regions) |
| Gas operations | One |
| Customer services | One |
| Vehicle fleet services | One |
| Supply chain | One |
| Information technology | One |
| Corporate real estate | One |
| Total | 24 |

[a]These organizations are part of Duke Energy's commercial (unregulated) business unit

# 7  Duke Energy: Engaging Employees in Sustainability

## Duke Energy's Maturity Curve Tool

The purpose of this tool is to help you develop a plan to begin moving the needle on sustainability within your department. To start, it helps to first think about the current state of sustainability within your organization. To help you do that, the graphic below illustrates at a very high level the different levels of sustainability maturity an organization can pass through over time.

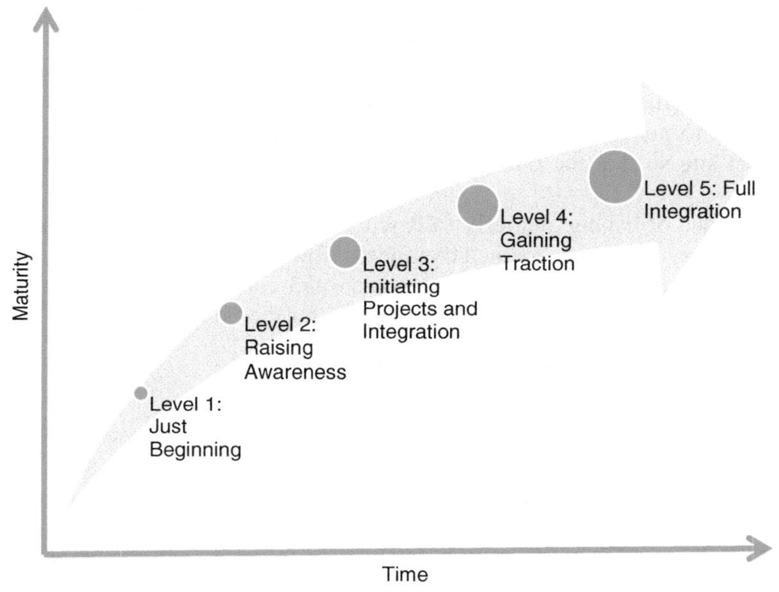

*Maturity vs. Time graph showing:*
- Level 1: Just Beginning
- Level 2: Raising Awareness
- Level 3: Initiating Projects and Integration
- Level 4: Gaining Traction
- Level 5: Full Integration

Based on the following descriptions, which level best describes the current state of sustainability within your organization?

☐ **Level 1: Just Beginning** - There is limited understanding of what sustainability means and why it is important. Efforts to put sustainability into practice are ad hoc or just beginning.

☐ **Level 2: Raising Awareness** - Efforts to raise awareness and support for doing business in a way that is good for people, the planet and profits are getting underway. Specific initiatives to help move the needle are starting to be considered.

☐ **Level 3: Initiating Projects and Integration** - Sustainability considerations are starting to be integrated into key projects and processes. Projects that are expected to deliver triple bottom line benefits are being initiated.

☐ **Level 4: Gaining Traction** - Efforts and projects are beginning to gain traction. The organization is starting to realize triple bottom line benefits.

☐ **Level 5: Full Integration** - Significant triple bottom line benefits are being achieved. Sustainability is being integrated throughout the organization to help drive innovative solutions.

**Fig. 7.8** Duke Energy's Maturity Curve Tool

management of the S Leads program, every member of the department (5 total) is asked to serve as a Coach to a subset of the S Leads to help spread the workload.

To better equip the S Leads, we also developed a new tool, the Maturity Curve Tool (Fig. 7.8), to help them think about the current state of sustainability in their departments. The S Leads were also provided a list of potential actions they could undertake, with support from their executives, to move the needle on sustainability in their departments. At the same time, they were encouraged to come up with new ideas that would resonate within their departments.

A simple template was also provided to the S Leads for them to document their plan for the coming year. This was more formal than the earlier approach of just asking them to pursue three key opportunities. We posted the annual plans on a SharePoint site so that the other S Leads (and Corps members) could reference them.

The practice of having quarterly calls with all of the S Leads (and Coaches) to share successes and lessons learned continues. Now that there are only 24 of them, each S Lead is expected to present their plan and current status at least once a year. This too has helped increase accountability for the S Lead role.

## 7.11 S Corps Program Enhancements

As was mentioned earlier, the S Corps is a Duke Energy's grassroots network of employees. Its mission is to further our progress as a best-in-class sustainable company. A review of feedback and survey results for the program over 2009–2012 indicated high levels of satisfaction with the program. In fact, over 80 % of the participants rated the workshop "extremely valuable" or "valuable."

The most substantial change we made to the program post-merger was the addition of "custom" and "onsite" workshops. These were added to our ongoing "traditional" offering. Each is described below:

- Traditional—The traditional workshop is open to employees across the company. To participate, employees travel as necessary to a video conference room in one of Duke Energy's major locations to attend the workshop. As part of the training, each employee is asked to commit to a PSP and undertake an improvement project in his/her workplace. Five-to-six traditional workshops are typically held annually.
- Custom—The custom workshop is tailored to a specific work group and typically sponsored by the work group's S Lead. The training is delivered via video conference or in a central location. Similar to the traditional workshop, employees are asked to commit to a PSP; however, the improvement projects are often undertaken by teams of four-to-five employees. To expose them to sustainability, management can also be asked to attend portions of the workshop and or a report-out meeting once the teams have developed their recommendations.

- Onsite—If a location recruits 10 or more employees, the sustainability department will bring the S Corps workshop to that location. This option was added because some employees at power plants and other field locations find it difficult to travel to a video conference room in one of Duke Energy's major locations. The employees do not have to be in the same work group. If the employees are largely in the same work group, the content is tailored to that group. Improvement projects can be undertaken individually or by teams of employees.

Similar to the S Leads program, while one member of the sustainability department provides overall management of the S Corps program, other members of the department provide support, specifically by taking a turn leading the workshops and coaching a class.

Another tweak that has been made to the program is the number and timing of the workshop follow-up teleconferences. Instead of three follow-up calls at 2, 5, and 12 weeks after the workshop, we now have two follow-up calls at 4 and 16 weeks after the workshop. We are also more clear about the milestones that should be accomplished by the time the follow-up call occurs. Specifically, by the first follow-up call in 4 weeks, S Corps members are expected to have their PSPs selected and entered into the online PSP Tracker. By the second and final follow-up call in 16 weeks, S Corps members are expected to have their work-related projects underway and entered into the online Project Tracker.

We also now send the S Corps members (and Leads) a brief, monthly electronic newsletter to keep them engaged and abreast of sustainability news and developments long after the workshop has been completed. They also get invited to optional events, such as continuing education sessions and volunteer opportunities. In several cases, S Corps members have ultimately been asked to serve as S Leads by their management.

## 7.12 Sustainability 101 eLearning Module

After the Progress Energy merger, Duke Energy grew approximately from 18,000 to 28,000 employees. To enable us to reach this increased number of employees more quickly, we created and launched a Sustainability 101 eLearning Module.

The sustainability department developed the Sustainability 101 eLearning module with the help of an outside design firm. The goal was to make it interactive and fun. We also partnered with Duke Energy's training department to finalize and deploy the eLearning module through the company's existing training system.

The front end of the module includes a brief introductory video on sustainability—what it means and why it is important to Duke Energy. The back end of the module includes "at home," "at the office," "at the power plant," and "in the field" scenarios. Participants enter these scenarios and then click to find and learn about real-life sustainability opportunities. This interactive, "gaming" feature was well received and several employees commented it was one the best online training modules they have ever completed at Duke Energy.

Even though the Sustainability 101 eLearning module was optional, 70 % of Duke Energy employees had completed the course as of April 2014. It is being provided to new employees as a standard part of the orientation process. We have made the eLearning module a prerequisite for the S Leads and Corps workshops. Because it provides a very robust introduction to sustainability, we have now been able to reduce the length of the S Leads and Corps workshops from 6 to 4 hours.

## 7.13 The Spark Fund

The Spark Fund is another post-merger addition to Duke Energy's employee engagement program. The purpose of the Spark Fund is to increase the number of sustainable projects delivering tangible results, support employee engagement, spur innovation, and visibly demonstrate the ROI of sustainability.

To create the Spark Fund, the sustainability department earmarks about $40,000 in its annual budget. Given the relatively small size of the fund, S Leads and Corps members are the only employees eligible to apply to the fund.

The Spark Fund provides seed funding for internal projects that have a strong sustainability business case, but would not otherwise be funded by the operational or corporate unit. S Leads and Corps members submit requests for project funding using a standard form that has been developed. Typically two waves of funding are awarded during the year and a selection committee comprised of a subset of the S Leads (those who have not submitted requests) select the "winning" projects (Fig. 7.9).

Sally Varner | Performance Excellence Leader, Outage and Maintenance Services, S Corps Member

"Thanks to help from the Spark Fund, we were able to replace inefficient, old-style lighting in our maintenance center that performs complex repairs to key equipment at our power plants. The new, energy efficient LED high bay light fixtures have provided multiple benefits: vastly improved lighting quality, less noise and electricity usage, lower O&M costs, avoided emissions and improved safety."

**Fig. 7.9** A Duke Energy Spark Fund Recipient

Requests for funding are evaluated, scored, and ranked based on the following criteria:

- Measureable, triple-bottom-line results—Projects that are expected to result in quantifiable cost savings, cost avoidance or new revenue streams and quantifiable environmental/social benefits.
- Innovation—Creative, new ideas that add value and are repeatable in other groups.
- Soundness—Projects that have been vetted with appropriate stakeholders and have a high probability of successful completion within the budget year.

The amount of money awarded for projects has varied from $1000 up to $15,000. Articles announcing the winners are posted on the Duke Energy intranet site. In addition, the funding recipient provides a final report to the sustainability department once the project is complete describing the results that were achieved.

## 7.14 Employee Engagement Results

In 2008, Duke Energy set out to develop an approach and set of internal programs to engage employees on sustainability. The original goal—to engage employees, foster innovation, and provide tangible business value to the company—has been accomplished. The original design of the approach was still sound, even after the merger with Progress Energy in 2012 substantially increased the size of the company.

The sustainability improvement projects that have been undertaken by Duke Energy employees have ranged from simple projects to reduce paper and postage to more complex projects at our power plants. Here are just a few examples:

- Three nuclear sites implemented filtered-water systems to replace bottled drinking water, saving approximately $180,000 annually and eliminating around 600,000 plastic bottles per year.
- One of our Midwest coal plants optimized cooling tower operations to reduce oil consumption saving $15,000 annually.
- Distribution design engineers in South Carolina worked to increase the use of mobile Internet devices when engineers go into the field to avoid multiple trips back to the office. This increases employee efficiency, reduces fuel and paper costs, and protects the environment by avoiding vehicle emissions.
- Our Corporate Real Estate team has implemented energy-saving projects that have saved an estimated $3 million over six years and helped us "walk our talk" on energy efficiency.
- Our Vehicle Fleet Services team deployed an Anti-Idling Policy that helps save an estimated $1.2 million annually in fuel, helps ensure compliance with emerging local/state regulations, and contributes to improved air quality.

- Our Claims Department changed the way they bill people who have damaged the company's property to cut the number of mailings in half. This simple change saves an estimated $70,000 annually in labor costs, postage, printing, and paper.

In addition to the types of tangible benefits described above, another result we have seen is more engaged employees. In fact, 80 % of those who joined the S Corps in 2014 indicated their pride in Duke Energy had increased as a result of the workshop. In addition, 100 % indicated they were "likely" or "very likely" to be better stewards of environmental, financial, and people resources in the future. Sustainability was also identified as a top-ten driver of employee engagement in the annual employee engagement survey that was sent to all Duke Energy employees in 2014. Specifically, employees who rated the company high on its sustainability performance tended to have higher overall engagement.

## 7.15 Lessons Learned

Using an experienced consultant to help us initially design our employee engagement approach back in 2008 was a good decision for Duke Energy. Given how early we were in our sustainability journey at the time, we no doubt saved time and avoided pitfalls by not starting from scratch ourselves. At the same time, a company's approach to engaging employees has to fit the culture and context of the company or it will not be successful—and only company staff can provide this perspective. Therefore, if a company chooses to use a consulting agency to help design its employee engagement program, it is essential for there to be a very close working relationship.

Building a large-scale program to engage employees can be daunting in the beginning. The key is to get the key planks in place to get the program going. Specifically, we first had to get executive support for the overall program. Once we had that, the executives provided us nominations for the S Leads. The S Leads in turn helped us get the S Corps program going.

Once employees are interested in sustainability, they need practical tools to help them identify and make improvements. Our original tools—the Energy Tool, Sustainability Filter, and Opportunity Finder—have served in this role. While they have been refined over the years, they have stood the test of time. We have also added new tools and resources—the Maturity Curve Tool, the Spark Fund and the Sustainability 101 eLearning module—to help enhance and further our efforts.

Going forward, we plan to continue to refine and build upon Duke Energy's employee engagement program. For example, an audience we hope to better reach in the future is middle management. We will strive to identify efficient and effective ways, such as integration into current communications and training, to better engage middle management. The goal is to spur new ideas and generate additional support for employee efforts.

In addition, now that the S Corps has grown to almost 400 employees, we hope to create regional chapters. While some ad hoc activities already exist regionally, we would like to see an S Corps member or two in each region take a more active role in identifying and implementing regional activities.

The fact that many employees already have very full plates is a challenge for any employee engagement program. This is especially true the first few years after a merger due to the large number of integration projects that have to be completed to bring the new company together. We have recognized this fact head on and encouraged the S Leads and Corps members to integrate sustainability into existing efforts and projects that are underway. This is a great way to get started, especially when time constraints will not permit the start of many new projects. This also reinforces our strong belief that sustainability should be baked into everything we do—and not be something separate that gets pursued as time allows.

## 7.16 Closing Comments

Engaging employees on sustainability is a great way to unlock creative ideas and innovation. It also increases employee pride in the company and helps with employee retention. In addition, Duke Energy's employee engagement program has provided the company with significant avoided costs, along with environmental and community benefits.

We encourage other companies who are just starting out to consider the approach Duke Energy is using to engage employees on sustainability and then tailor it to fit their culture. Companies who are already on this journey may just want to focus on certain aspects of our program for potential application. We have built our approach and supporting tools over time and hope that the information contained in this chapter will be of assistance to other companies.

Now more than ever, companies need to get all good ideas on deck. The more our employees learn about sustainability and put it into practice, the more successful we will all be in the future. Employee engagement is a journey that is definitely worth taking.

# Chapter 8
# Entergy: Climate Change Resiliency and Adaptation

**Brent Dorsey**

**Abstract** Effectively managing environmental risks is essential to Entergy's ability to create long-term, sustainable value for owners, customers, employees, and communities. Given the utility's first-hand experience with hurricanes, storm surges, and a disappearing coastline, Entergy leaders have strived to gain an understanding of short-term and long-term implications to the region from climate change. Entergy has put its high-level commitments into very specific actions by driving major initiatives to address climate-change adaptation and resilience from a regional perspective. Entergy also is working to improve the resilience of its generation, transmission, and distribution infrastructure. An example of an infrastructure improvement project is the rebuilding of Entergy's gas system in New Orleans following Hurricane Katrina—one of the world's largest gas-rebuild effort in history resulting from a catastrophic event. Entergy believes that stakeholder outreach is instrumental in developing effective strategies to protect and strengthen infrastructure. In collaboration with local universities, Entergy has hosted technical conferences with customers to learn how to prioritize infrastructure investments in ways that align with actions being taken at local levels. These and other outreach efforts demonstrate the importance of working collectively with stakeholders to build resilient communities in a responsible, sustainable, and cost-effective manner.

---

B. Dorsey (✉)
Corporate Environmental Programs, Entergy, New Orleans, USA
e-mail: bdorsey2015@outlook.com

## 8.1 Our Risks Down on the Bayou

As residents prepared for a relaxing season of music and crawfish for the summer of 2005, unrelenting weather was brewing off the Gulf Coast. That year's storm season proved to be the most active Atlantic/Gulf of Mexico hurricane season on record. The National Weather Service even ran out of storm names, resorting to the use of Greek names (Alpha–Gamma) for the last seven storms of the season.

Even today, images of damage caused by hurricanes Katrina and Rita are burned into the mind's eye of people across the country, most especially, residents of the Gulf Coast. Katrina caused loss of power for over 1.1 million people and destroyed 275,000 homes. Rita put nearly 800,000 people in the dark while another 175,000 in Texas dealt with rolling blackouts.

In the Gulf Coast region, where a large portion of Entergy's customer base and the majority of its utility infrastructure are located, serious environmental, social, and economic consequences are resulting from the effects of climate change. Extreme weather events of 2005 revealed that no longer was a build and rebuild approach for storm management viable. The storms and their aftermath provided a clear business case for Entergy to forecast and mitigate climate-related risks (Fig. 8.1).

Coastal Louisiana suffers one of the fastest rates of wetlands loss in the world, and restoration costs are estimated in the tens to hundreds of billions of dollars. According to the U.S. Geological Survey, an average of 34 square miles of south Louisiana land, mostly marsh, has disappeared each year for the past five decades. From 1932 to 2000, the state lost 1,900 square miles of land to the Gulf of Mexico, an area roughly the size of Delaware. By 2050, if nothing is done to stop this process, the state could lose another 700 square miles, and a third of the 1930s coastal Louisiana will have vanished (Fig. 8.2). New Orleans and surrounding areas will become more vulnerable to future storms through loss of this natural storm surge buffer (Fig. 8.3).

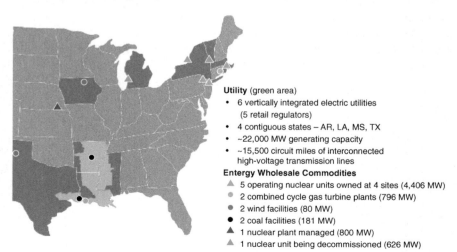

**Fig. 8.1** Entergy Corporation operating areas (2015)

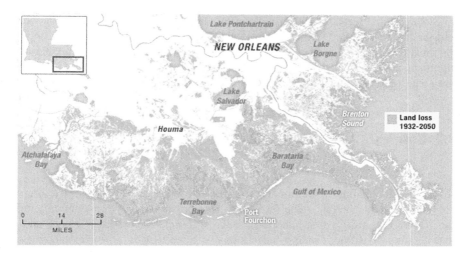

**Fig. 8.2** Projected land loss through 2050. *Source* Louisiana Coastal Wetlands Planning Protection and Restoration Act Program

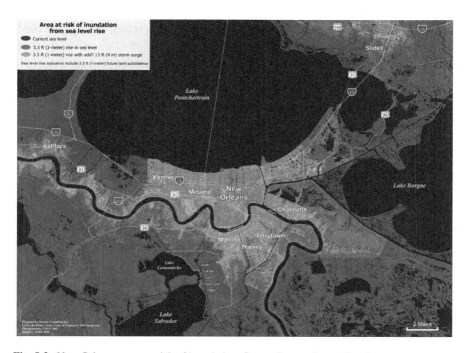

**Fig. 8.3** New Orleans areas at risk of inundation. *Source* Stratus Consulting, Inc.

Since 2005, a series of catastrophic events marked by hurricanes Katrina, Rita, and Ike, as well as the Deepwater Horizon oil spill, have further devastated one of the most fragile landscapes on the planet. Such events also highlighted the region's significance and exposed its vulnerabilities. In this rapidly changing physical environment, industries and communities must be resilient to survive.

## 8.2 Entergy's Approach to Climate Change, a Legacy of Stewardship

Effectively managing environmental risks is essential to Entergy's ability to create long-term, sustainable value for owners, customers, employees, and communities. Given the utility's first-hand experience with hurricanes, storm surges, and a disappearing coastline, Entergy leaders have strived to gain an understanding of short-term and long-term implications to the region from climate change. Driven by real risks and exposure, Entergy's challenges in addressing climate-change resiliency and adaptation are virtually unrivaled in the U.S. utility industry.

Entergy has put its high-level commitments into very specific actions by driving major initiatives to address climate-change adaptation and resilience regionally. Entergy helped fund a landmark adaptation study presented during America's Energy Coast policy forum in 2010 that was the first comprehensive analysis of climate risks and adaptation economics for the Gulf Coast region. In 2012, Entergy partnered with America's WETLAND Foundation on the Blue Ribbon Resilient Communities initiative, which helped vulnerable communities identify key parameters for proactive and reactive dimensions to storm readiness and recovery.

Entergy also is working to improve the resilience of its generation, transmission, and distribution infrastructure. One example of infrastructure improvement is the rebuilding of Entergy's gas system in New Orleans—one of the world's largest gas-rebuild effort in history resulting from a catastrophic event. After the gas system was flooded by saltwater from Hurricane Katrina, Entergy began converting 844 miles of piping in the city's low-pressure system to a high-pressure system made of high-density polyethylene, which is virtually impervious to saltwater corrosion. In 2009, it was named Platts' Global Infrastructure Project of the Year.

Entergy believes that stakeholder outreach is instrumental in developing effective strategies to protect and strengthen infrastructure. In collaboration with local universities, Entergy has hosted technical conferences with customers to better prioritize infrastructure investments in ways that align with actions being taken at local levels. These and other outreach efforts demonstrate the importance of working collectively with stakeholders to build resilient communities in a responsible, sustainable, and cost-effective manner.

Entergy's regional efforts to address climate-change resiliency and adaptation have been a natural evolution for the company. In 2001, Entergy became the first U.S. electric company to publicly announce a voluntary greenhouse gas emission target and to develop long-term targets to help combat climate change.

The company's commitment was initiated by the leadership of its board of directors and then-Chief Executive Officer J. Wayne Leonard. Entergy was the first U.S. electric company to publicly establish such a greenhouse gas emission target. To implement this target, the New Orleans-based company partnered with Environmental Defense, a national advocacy group, to develop a program to reduce carbon dioxide emissions from Entergy's plants that generate electricity through burning fossil fuels. Entergy was the first U.S. electric company accepted for membership in the Partnership for Climate Action (PCA), a collaboration of international business and environmental leaders dedicated to climate protection established by Environmental Defense.

"Limiting greenhouse gas emissions as a company, as a nation, and globally had to start somewhere. Entergy's management decided that this company could provide that starting point," said Chuck Barlow, Entergy's vice president of environmental strategy and policy. "Neither our company nor our industry can solve what will be a global issue for decades to come, but we can provide an example of responsible steps in the right direction."

To support its voluntary GHG emission goal in 2001, Entergy dedicated $25 million in supplemental funding through a new Environmental Initiatives Fund (EIF). Five years later, after exceeding its stabilization commitment and reducing greenhouse gas emissions by 23 percent, Entergy made a second GHG commitment. In 2011, Entergy's third commitment, based on principles embodied in its Environment$^{2020}$ strategy, became effective. Since 2001, Entergy's EIF has purchased more than $6 million in carbon dioxide equivalents ($CO_2e$), or GHG offsets.

In tandem with renewed commitments to reduce greenhouse gas emissions, considerable support through the EIF continued. At the end of the first commitment in 2005, emissions had been reduced 21 percent while sales had increased 23 percent. The majority of the sales increase came from the low-GHG emitting merchant nuclear fleet, producing a good result. Management needed assurance that the second and third reduction commitments could be accomplished without putting the company at risk. To ensure this, GHG offset banks purchased during the first commitment were used to cover any overages. In addition, the "head room" gained from the first commitment was also applied. This head room reflected the savings between the actual emissions and the commitment to stabilize emissions at year 2000 levels during the period 2000–2005.

Ultimately, the EIF was funded at $1 million per year to focus on community environmental improvement projects. Internal projects would still be contemplated but funded through other means.

In 2002, Entergy's board of directors articulated the company's commitment to the environment with adoption of an environmental vision statement that sets expectations in areas of sustainable development, performance excellence, and environmental advocacy. In 2011, Entergy updated its commitment with the Environment$^{2020}$ plan, a six-plank strategy emphasizing proactive adaptation measures to mitigate physical risks to Entergy's operating area. This was necessary not only to minimize overall business-interruption losses from extreme weather, but also to preserve economic prosperity of local communities and protect the safety of customers.

Operating business in an environmentally responsible way produces environmental, economic, and social benefits for all of Entergy's stakeholders. For owners, it helps avoid costs associated with noncompliance and mitigates business risks posed by climate change and other environmental issues. For customers, environmentally sound operations protect public health and safety, and energy-efficiency efforts result in lower electricity usage. For communities, Entergy's adaptation measures help protect water, air, and biodiversity and improve quality of life. For employees, Entergy encourages direct involvement by sponsoring volunteer opportunities that support environmental and community-improvement projects. The company is focused on developing and implementing charitable giving, volunteerism, and low-income customer service strategies that position both the company and communities for sustainable growth (Fig. 8.4).

Managing risks of climate change involves anticipating regulatory and physical risks, testing business decisions against scenarios of potential change, identifying where Entergy is vulnerable, and devising sound, cost-effective business strategies to manage risks and recognize opportunities to prosper in a changing world. This includes processes that address business continuity, storm-recovery readiness, and storm hardening to prioritize investments that reduce business-interruption losses. Storm hardening is the process of design and implementation of new or retrofitted infrastructure such as transmission and distribution facilities (lines, substations, poles, and conductors) capable of withstanding more extreme weather events. Managing risk also includes stakeholder outreach to ensure that resiliencyinvestments complement actions customers and communities are taking to uphold prosperity, safety, and quality of life. Entergy's management approach to address and adapt to environmental risks includes engaging with regional, state, and local governments, universities, nongovernmental organizations, and businesses that share similar interests in building resilience.

**Fig. 8.4** More than 100 Entergy volunteers helped launch floating islands in a demonstration project of a new technology to protect the Gulf Coast south of Houma, La., considered to be ground zero for coastal land loss in America. The floating islands are man-made ecosystems anchored in place in open water that mimic naturally occurring wetlands. They provide nature a starting point to help jump start wetland restoration.

## 8.3 Adaptation Study: Building a Resilient Gulf Coast

Following devastating hurricanes in 2005 and 2008, Entergy recognized the need to focus not only on business continuity, but also on prosperity and resilience for communities. The company's $1.5 billion loss as a result of hurricanes Katrina and Rita in 2005 reflects only a fraction of the $150 billion loss that communities suffered from Katrina alone.

Entergy set out to develop a comprehensive, objective, consistent fact base to quantify climate risks in the Gulf Coast region and inform economically sensible approaches for addressing this risk. In 2010, Entergy and America's WETLAND Foundation released a landmark study that estimated the physical and financial risks to the energy coast of the Gulf of Mexico, as well as the impacts of sea level rise, subsidence, storm surge, and wind impacts for 77 counties and parishes in Texas, Louisiana, Mississippi, and Alabama at the zip code level. It also identified adaptive measures that could help reduce growth in these risks on a sector-by-sector basis.

Data from the $4.2 million adaptation study quantified the economic value at stake: the livelihoods of 12 million people; natural resources that support $634 billion in annual gross domestic product; and assets and critical infrastructure valued at more than $2 trillion that are increasingly vulnerable to storm surge, flooding, wind damage, and the effects of sea level rise (Fig. 8.5).

The adaptation study was presented during America's Energy Coast policy forum in New Orleans at the conclusion of the DELTAS2010: World Deltas Dialogue (http://www.deltas2010.com) in October 2010. It showed that without

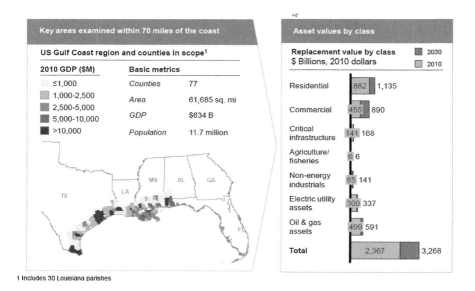

1 Includes 30 Louisiana parishes

**Fig. 8.5** Adaptation study area. *Source* ESRI, energy velocity

large-scale adaptation efforts, the region's gross domestic product may be unable to keep up with costs of protecting nationally critical assets. These critical assets include oil and gas pipeline networks, refineries, chemical plants from the Gulf of Mexico and production within the Gulf South region, as well as transportation hubs of the ports, intercostal waterways. This study was greatly strengthened and enriched by contributions from many participants, including Swiss Re, which brought its natural catastrophe and climate risk assessment research and knowledge to bear on the challenge of quantifying climate risks. Methodology used in the study was devised and tested by a consortium of public and private partners who helped develop a factual framework for decision makers to build a portfolio of economically suitable adaptation measures.

Addressing assets in 23 different classes (commercial, residential, industrial, and infrastructure), the study included a detailed assessment of more than 500,000 miles of electric transmission and distribution assets and approximately 300 generation facilities. Three scenarios were considered for 2030 and 2050 representing low, average, and extreme climate change. Results showed that the Gulf Coast region faces $14 billion in average annual asset losses from today's climate, increasing to $23 billion with the high scenario. Cumulative losses over the next 20 years likely will exceed $350 billion, representing two to three percent of the region's gross domestic product. Extreme-loss years may get worse and occur more frequently.

Fifty different adaptation measures were evaluated to determine their applicability, cost, and the estimated asset loss that could be avoided. This work identified a set of potentially attractive measures that can address almost all of the increase in loss going forward, including actions to mitigate an approximate $7 billion per year in annual expected loss in the 2030 timeframe. Measures translate to nine broad efforts to reduce risk across all sectors:

1. Improved building codes,
2. Beach nourishment,
3. Wetlands restoration,
4. Levee systems,
5. Improved standards for offshore platforms,
6. Floating production systems,
7. Replacing semi-subs with drill ships,
8. Levees for refineries and petrochemical plants,
9. Improving resilience of electric utility systems.

The project assembled a supply curve of adaptive measures for consideration within the study area. As shown in Fig. 8.6, these measures were expressed in terms of their cost/benefit reductions to casualty loss.

For example, the first measure, "resilience, new distribution," has a cost/benefit of 0.17, which means that for an investment of $0.17, one can expect a reduction in casualty loss of $1. If we accept all measures that are less than or equal to 1.0, the cumulative loss reduction would be between $4.5 billion and $5 billion per year. The one-to-one breakeven point on the supply curve falls between "higher

## 8 Entergy: Climate Change Resiliency and Adaptation

**Potentially attractive measures can address the increase in annual loss between today and 2030 and keep the risk profile of the region constant**

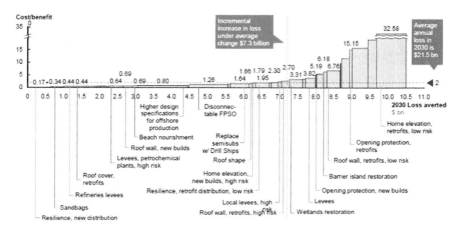

Fig. 8.6 Supply curve of adaptive measures

design specifications for offshore production" and "disconnectable FPSO" (floating production storage and offloading).

Co-benefits of measures were not included in the initial analysis, but could be used to provide additional benefits for comparison to costs. In other words, if the cost/benefit level is expanded to 2.0, loss reduction would be almost $7 billion per year. Cost/benefit is derived by looking strictly at reduction in casualty loss. If the scope were broadened to include other benefits beyond casualty loss, more measures would achieve cost/benefit parity. For example, "wetlands restoration" produces a cost/benefit of 3.31, which means $3.31 would be spent to reduce casualty loss by $1. But when other co-benefits are added to the equation, such as increased biodiversity or wetlands-based revenue activities (eco-tourism, hunting, etc.), then wetlands restoration would come closer to clearing economic hurdles (Fig. 8.7).

Similarly, a subset of the overall supply curve of adaptive measures was assembled for utility-specific alternatives (Fig. 8.8). These measures have been estimated for utilities within the four-state, 77-county/parish study area and are not specific to Entergy. If the industry were to invest in each cost-effective measure, one could expect a reduction of casualty loss of more than $800 million.

The study estimated that public funding of $44 billion will be required over the next 20 years for key infrastructure projects, including wetlands and levees. Some $76 billion in private funding will also be required. Policy makers may need to support and offer incentives for private capital investment, such as subsidizing homes built to meet higher building codes in low-income areas.

The primary focus was on assessing measures that are known and executable today instead of assessing future innovations in technology. This choice has the

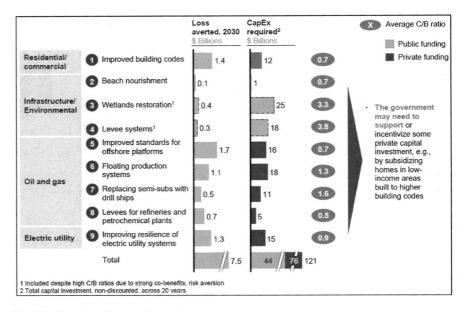

**Fig. 8.7** Cost/benefit ratios of adaptive measures

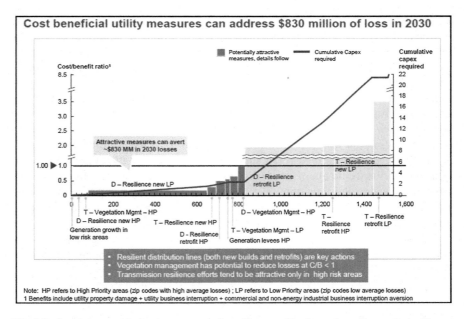

**Fig. 8.8** Supply curve of adaptive measures for utility-specific alternatives. *Source* Swiss Re

benefit of making measures more tangible to decision makers, but results of the analyses must be considered from the point of view that future developments (i.e., new building materials or methods that may be more hurricane-resilient) can be incorporated. Correspondingly, the analyses should be repeated periodically—ideally every five years—to incorporate new innovations in adaptation.

Read the adaptation study report: http://www.entergy.com/content/our_community/environment/GulfCoastAdaptation/report.pdf.

Read the executive report: http://www.entergy.com/content/our_community/environment/GulfCoastAdaptation/Building_a_Resilient_Gulf_Coast.pdf.

## 8.4 Blue Ribbon Resilient Communities Initiative

As an extension of the adaptation study, in 2011, Entergy and America's WETLAND Foundation (AWF) launched 11 outreach forums in coastal communities addressed in the study. Spanning 14 months and five states, the forums brought together more than 1,100 leaders and community representatives for a dialogue on local coastal issues and specific vulnerabilities.

Forums identified eight key parameters for proactive and reactive dimensions to storm readiness and recovery, known as "Combined Resiliency Indexes:"

1. Governance,
2. Society and economy,
3. Coastal restoration and protection,
4. Land use and structural design,
5. Risk knowledge,
6. Warning and evacuation,
7. Disaster recovery and emergency response,
8. Transparency, public education, and awareness.

In advance of each Blue Ribbon forum, AWF conducted community research through local focus groups and interviews. Respondents were asked to discuss their community's values and to rate their community's performance on a number of resiliency indicators. Combined responses generated a resiliency index for each community. Entergy contributed results of a study on Gulf Coast resiliency and sustainability, quantifying the economic value of what was at stake for each community and establishing the magnitude of risk.

Figure 8.9 is a spider-web graph summary of combined resiliency indexes with detail from each forum. If a community were assessed as fully prepared and resilient, it would receive a perfect score of "5" for each of the eight dimensions. This analysis provided a simple, easily understood graphic approach to evaluate each community, with strengths and weaknesses helping identify which gaps are most prevalent.

These layers of local feedback, combined with detailed research, gave forum participants a groundbreaking opportunity to assess their communities' vulnerabilities and outline steps to improve resiliency. Input gathered over the course of

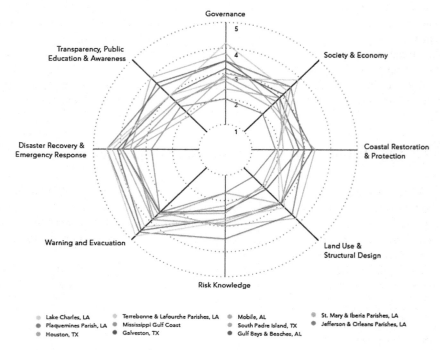

**Fig. 8.9** Spider-web graph of combined utility indexes

the forums came from policy makers, agency officials, business leaders, concerned citizens, and the gamut of professionals on the front lines—environmental managers from all fields, engineers, energy sector workers, farmers, fishermen, public works officials, and community activists, among others.

Participants' insights generated dozens of recommendations impacting state and federal policies. The most important impact of the Blue Ribbon initiative, however, was community empowerment. Forums helped mobilize the collective energy, expertise, and long-term planning and problem-solving skills of an entire region to protect its heritage and secure its future.

Working in collaboration with partners, Entergy continues to assess environmental risks, identify possible solutions, and make adaptation a high-priority local issue.

## 8.5 Blue Ribbon Congressional Briefing in Washington, D.C.

In September 2012, AWF and its partners, including Entergy, presented the final Blue Ribbon Resilient Communities report to legislators on Capitol Hill.

**Fig. 8.10** Sponsors of the report including Chuck Barlow, Entergy's Vice President of environmental strategy and policy, presented the final Blue Ribbon Resilient Communities report at a Capitol Hill briefing on September 12, 2012

"Beyond Unintended Consequences" provided 30 recommendations for Gulf Coast adaptation and was supported by congressional leadership from the Gulf Coast region. The presentation also highlighted Louisiana's Comprehensive Master Plan for a Sustainable Coast (http://www.lacpra.org/index.cfm?md=pagebuilder&tmp=home&nid=24&pnid=0&pid=28&fmid=0&catid=0&elid=0) (Fig. 8.10).

The Louisiana master plan was developed and approved by the state legislature and itemizes $50 billion in investments over 50 years aimed at restoring wetlands habitat and building flood protection through restoration of natural buffers (marshes and barrier islands) to hurricanes and storm surges.

Actions included in the Louisiana master plan represent a far-sighted and proactive solution that will help safeguard the citizens and industries of the Gulf Coast from consequences of human actions, both historic and ongoing. The plan also provides economic opportunity for thousands through job creation and development investments, all while strengthening communities.

Read the final Blue Ribbon report: http://futureofthegulfcoast.org/AmericasWETLANDFoundation_Beyond.pdf.

## 8.6 Coastal Resilience Technical Forums

Partnering with Louisiana State University and Lamar University, Entergy organized two technical conferences to extend the coastal resilience dialogue to academic and industry audiences. Conferences were independent of Entergy's work with AWF, but were regarded as complementary efforts to strengthen community engagement.

Participants' input was sought to help prioritize system-hardening investments to minimize economic impacts from business-interruption losses. Entergy developed the conference agenda from data gathered during customer interviews about perceived vulnerabilities, actions taken to improve resiliency, and expectations of their utility service providers.

Entergy asked participants to focus on storm-hardening strategies in two areas in Louisiana and Texas and proposed three additional phases of system hardening that would take place over the next ten years. In addition to estimating the hardening-strategy cost, participants used the Gulf Coast adaptation study to estimate avoided-loss benefits to the area from hardening investments.

Economists from the University of Texas and Louisiana State University then estimated the extent of economic losses to the region from projected wind damage, flooding, and storm surge, emphasizing how losses rippled through the economy. They also estimated the benefit in avoided losses that would result from the pilot hardening initiatives. In both pilot cases, independent economists estimated the benefit-to-cost ratio for hardening investments at 5 to 1, meaning each dollar invested in hardening assets would produce five dollars' worth of benefits to the regional economy.

## 8.7 Building Coalitions, Driving Innovation for the Future

Entergy's adaptation and resiliency efforts also leverage the funding of projects from the shareholder-funded EIF program, which seeks partnership projects to provide registered greenhouse gas offsets, develop innovative solutions to climate-change impacts, and support coastal and wetlands restoration. The EIF identifies projects that help support greenhouse gas reductions, wetland restoration, adaptation, and other projects aimed to reduce environmental footprint and reduce physical risks.

Through the EIF, Entergy funded development of the first carbon-offset methodology for emission reductions from deltaic wetlands restoration, which was approved for use by the American Carbon Registry in 2012. Developed by Tierra Resources, the carbon-offset methodology tool created a potentially self-sustaining revenue source for wetlands restoration through the sale of carbon offsets. When Mississippi River delta wetlands are restored, landowners can use the methodology to calculate the amount of greenhouse gases the rebuilt wetlands will absorb over time. The result is registered carbon credits, which landowners can sell to companies seeking to offset their greenhouse gas emissions. Proceeds from the sale help offset landowners' cost of wetlands restoration. In 2013, Entergy received an Innovation Award from American Carbon Registry for this work.

Entergy provided additional funding to pilot the first wetlands restoration offset project in the nation applying American Carbon Registry's methodology. The Luling Oxidation Pond Wetlands Assimilation project, 19 miles west of New Orleans discharges treated municipal wastewater into an adjacent 950-acre wetland property to help restore the wetlands' function and increase carbon sequestration.

Also in 2012, Entergy completed registration of a reforestation project in Arkansas and Louisiana that will remove an estimated 460,000 tons of carbon dioxide from the atmosphere over the next 40 years. The project restored nearly

3,000 acres of marginal agricultural land to native bottomland hardwood forests, which are forested wetlands that originally covered more than 30 million acres in the Lower Mississippi Valley. Most of these forests were destroyed by logging in the early 1900s and further reduced by conversion to agriculture in the 1960s and 1970s. Reforested lands were replanted with native species, primarily bald cypress and bottomland oaks in the Tensas, Red River, Overflow, and Pond Creek National Wildlife Refuges. By registering the reforestation project with the American Carbon Registry, Entergy further strengthened its ability to operate in a carbon-constrained environment using innovative, market-based approaches.

Other projects supported by the EIF include the Coastal Bayou Segnette Cypress Reforestation project (in partnership with Restore America's Estuaries, the Coalition to Restore Coastal Louisiana, and Jefferson Parish, La.), a marsh-restoration terracing project near Lafitte, La. (in partnership with Ducks Unlimited), and a pilot program of 17 electric vehicle-charging stations on four Texas college campuses and elsewhere in Entergy's service territory. Photos of the Bayou Segnette and Lafitte projects are in Figs. 8.11, 8.12 and 8.13.

**Fig. 8.11** The Bayou Segnette Cypress Reforestation project in the Barataria Preserve included construction of protective rock jetties to reduce erosion

**Fig. 8.12** Water hyacinths planted in Bayou Segnette provide natural marsh restoration

**Fig. 8.13** A year after terraces were built to restore marshland near Lafitte, La., submerged aquatic vegetation (*inset*) had begun to flourish

In 2013, Entergy identified and funded three additional projects with support from the EIF. Two are registered with the American Carbon Registry—IdleAir truck stop electrification technology and a low-methane rice farming system. The third is a marsh-restoration project that also provides carbon-offset opportunities.

In Mississippi, a truck stop equipped with IdleAir technology allows truck drivers to turn off their engines while maintaining a comfortable cabin temperature and enjoying the use of television, internet, and other electronics. In addition to reducing carbon emissions, IdleAir helps fleet operators reduce costs by conserving fuel and reducing engine wear. Entergy funded the opening of an IdleAir site in Pearl, Miss., with other potential sites in Arkansas and Texas.

In Arkansas, Entergy partners with Terra Global Capital, White River Irrigation District, and USDA Agricultural Research Service to test new rice-cultivation practices that can reduce methane emissions while providing savings to farmers and ecological benefits.

In Louisiana, Entergy supports marsh restoration in the Chef Menteur Pass Wetland Mitigation Bank in Orleans Parish. The Chef Menteur Pass property provides critical ecological and risk-reduction functions for the Gulf Coast of Louisiana and was identified in Louisiana's 2012 Coastal Master Plan as a priority for the state's marshland restoration activities.

Entergy also has created opportunities for people to get involved on an individual level. In 2009, Entergy and the Pew Center on Global Climate Change launched the Make an Impact program to help individuals take action to reduce their carbon footprint. In recognition of Earth Day the following year, Entergy launched Double Your Difference, an initiative that allowed individuals to purchase high-quality carbon offsets and double the impact of their purchases through Entergy's dollar-for-dollar match. By signing up for the program, participants could support specific projects to reduce greenhouse gases while maximizing their own efforts to go carbon neutral.

## 8.8 Lessons and Transferability

Entergy's approach to resiliency and adaptation focuses on sea level rise, subsidence, storm surge, and wind—hazards with the greatest potential for occurrence in the analysis study area based on historical data.

For the analysis to be transferred to other entities, such as utilities, industries, or communities, these entities must first identify primary hazards and risks for their study areas. These could include droughts and the associated fires or famines they can induce, floods, volatile changes in weather, extreme cold, ice and snow, mud slides, and avalanches, heat waves, tornados and other extreme weather events. Once hazards are established, an analysis including estimated impacts and benefits can be conducted. Clearly, each application of an analysis will involve different sets of hazards, measures, and impacts.

In its analysis and application, Entergy recognizes several opportunities for expansion. As previously observed, the analysis could estimate not only the direct casualty loss reductions provided by investments in measures, but also co-benefits. The example in Sect. 8.3 noted co-benefits of wetlands restoration in enhancing biodiversity and recreational opportunities. Expanding the analysis could modify benefits significantly for each measure and yield a more comprehensive cost/benefit ratio.

The Blue Ribbon Resilient Communities meetings provided interesting insights into public perceptions regarding perceived vulnerabilities and potential impacts of climate change. For example, many Texas participants thought of Texas as a state with a coast, but not a coastal state. Many Houston residents did not consider their home a coastal city, despite the fact that Houston's port is the busiest in the United States in terms of foreign tonnage and second busiest in overall tonnage. Had Hurricane Ike made landfall 30 miles farther west in 2008, the storm surge likely would have flooded downtown Houston and resulted in even more catastrophic damage and loss for Texas.

Community meetings also shed some light on attitudes regarding climate change and its political implications, including resistance that may be encountered when attempting to build the case for public investment in proactive adaptation strategies.

Technical conferences were equally productive and enlightening for participants. Entergy's environmental lead team became more aware of the need to establish partnerships with key industrial customers to prepare for weather-related hazards. Comparing service-restoration activities after hurricanes Rita and Ike revealed the range of damage scenarios that can result from different storms. Hurricane Rita was primarily a wind event that significantly impacted Entergy's transmission system. Repairs and service restoration for customers lasted several weeks. In contrast, Ike was primarily a flooding event that did not seriously damage Entergy's system; however, flood waters caused significant damage to industrial customers' facilities. When Entergy was ready to restore power, many of these customers were not able to receive service. This example demonstrates the need for both suppliers and customers to be more aligned on strategies and tactics to improve adaptation and resilience.

The company's support of the adaptation study allowed Entergy to communicate the impacts of a changing climate in terms that communities could understand and implement as actions. By focusing on resiliency, the company has effectively engaged its stakeholders on a new meaningful and important industry and community issue.

The EIF provides support to community adaptation and resiliency projects and programs by purchasing offsets and investing in wetlands restoration and reforestation. What started as a leap of faith has resulted in significant improvements for the company and its communities.

## 8.9 The Next Decade and Beyond

In addressing the effects of climate change, the utility industry is at a crossroads, with adaptation and resiliency at the center of the industry's potential transformation.

The traditional utility paradigm of central station generation has one set of adaptation and resiliency measures, while new and evolving utility constructs (e.g., smart grid, micro grid, distributed energy resources, renewables, new electrification technologies) may have to rely on alternative adaptive measures.

In addition to the electric system, regulation also is evolving and adapting to changes. Regulatory systems will need to identify and develop rate structures to recover adaptation and resiliency costs. Utilities and regulators should continue efforts to establish expectations for reliability, resilience, and adaptation, as well as other evolutionary transformations underway.

"Entergy's service area has experienced the impacts of coastal erosion, wetlands loss, and storm surge," said Barlow. "It really doesn't matter whether you believe that these impacts are from climate change or just mother nature at her worst; you can see the land loss, and you know there will be a future hurricane. The Gulf South is identifying adaptive measures to cope with these impacts, and Entergy will continue to be a leading part of that effort, whether you are talking about protecting the environment or protecting the electric grid. Both are important to our stakeholders. Both are important to Entergy."

For more information on Entergy's environmental initiatives and climate-change resiliency and adaptation efforts, visit. http://www.entergy.com/environment

# Chapter 9
# Exelon Corporation: Strategic Greenhouse Gas Management

Christopher D. Gould, William J. Brady, Melanie Dickersbach, Bruce Alexander and Alfred Picardi

**Abstract** Exelon Corporation has been a leading provider of reliable, low carbon generation since its formation in 2001. The company has advocated for a price on carbon and voiced support for climate change action. In 2008, Exelon announced its *Exelon 2020* goal to abate 15.7 million metric tons of greenhouse gas (GHG) emissions annually by 2020. By focusing on internal operations, overall grid emissions and customer energy efficiency programs, Exelon 2020 pushed the bounds of traditional GHG management by using an end-to-end value chain perspective. This strategic approach created value by informing and guiding investment decisions, shaping corporate culture, providing competitive advantage for customers and positioning the company as a leader on environmental public policy. The future of carbon policy remains uncertain, but by most indications, we are transitioning to a less carbon intensive world. Positioning the company for this future is fundamental to the sustainability of the business. The *Exelon 2020* program helped to bridge the topic and shows how reliability, customer and shareholder value, and competitive markets are an important part of the carbon conversation.

## 9.1 Introduction

How does a company with one of the smallest carbon footprints in its sector create value for customers, shareholders, and other stakeholders when the cost of carbon emissions is not accounted for in the price of its product? Exelon Corporation has

---

C.D. Gould · W.J. Brady · M. Dickersbach (✉) · B. Alexander · A. Picardi
Exelon Corporation, Chicago, USA
e-mail: Melanie.Dickersbach@exeloncorp.com

been a leading provider of reliable, low carbon generation since its formation in 2001. The company has advocated for a price on carbon and voiced support for climate change action. In 2008, Exelon announced its *Exelon 2020* goal to abate 15.7 million metric tons of greenhouse gas[1] (GHG) emissions annually by 2020. By focusing on internal operations, overall grid emissions and customer energy efficiency programs, *Exelon 2020* pushed the bounds of traditional GHG management by using an end-to-end value chain perspective. Despite failed federal legislation on carbon emissions (carbon[2]), a stagnant national economy and a sizable merger with Constellation Energy that resulted in an updated goal of abating 17.5 million metric tons of GHG emissions, Exelon achieved its goal in 2013, seven years ahead of schedule.

This strategic approach created value by informing and guiding investment decisions, shaping corporate culture, providing competitive advantage for customers, and positioning the company as a leader on environmental public policy. The future of carbon policy remains uncertain, but by most indications, we are transitioning to a less carbon-intensive world. Positioning the company for this future is fundamental to the sustainability of the business. Our *Exelon 2020* program helped to bridge the topic and show how reliability, customer and shareholder value, and competitive markets are an important part of the carbon conversation.

## 9.2 Exelon's Early Action

Exelon's initial strategy focused on positioning the company for a carbon constrained future through early action and a low carbon intensity profile.[3] Exelon's nuclear electric generating assets, which have no direct carbon emissions, have produced the majority of the company's earnings since its inception in 2001. A price on carbon would give the company's nuclear fleet a substantial competitive advantage over electricity generated with fossil fuel in the marketplace. Exelon's nuclear fleet would benefit from the potential power price uplift from the consequences of the regulation, while not incurring the compliance costs that its competitors' higher emitting generation sources might face. This was among the driving factors for Exelon to take a leadership position on the issue of GHG emissions pricing and policy. And while sometimes challenged on the environmental front by stakeholders, the low carbon emissions and 24/7 reliability of nuclear generation highlighted the important contribution of this type of generation to the issue of greenhouse gas mitigation and the ultimate transition to a low-carbon economy.

---

[1] Carbon dioxide, methane, and other compounds that trap heat in the atmosphere known collectively as "greenhouse gases" and referred to here as "GHG."

[2] Carbon—carbon is used in this chapter interchangeably with greenhouse gas, and referred to simply as carbon or carbon emissions.

[3] "Carbon intensity profile" means greenhouse gas emissions per unit of electrical generation, or "intensity."

9 Exelon Corporation: Strategic Greenhouse Gas Management

**Fig. 9.1** Excerpt from *Exelon 2020*: A Low Carbon Road Map, 2009 update

As we sought to develop our GHG program, Exelon was an early adopter under the EPA Climate Leaders Program and set our first GHG goal in 2005. After careful consideration, our goal was set for reducing absolute GHG emissions by 8 % from our 2001 levels by 2008. When set, we knew reaching this goal could be difficult, in part because 95 % of our emissions were tied to electric generation driven by customer demand.[4] We were also still developing our GHG accounting programs and learning where reduction opportunities might be available.

Some of the most difficult work was developing Exelon's GHG inventory. This included mapping out the sources at our then-37 power generation stations, two utility distribution systems, and supporting commercial buildings. In addition to identifying all the sources, we also needed to establish data requirements, contacts to obtain data, an inventory management plan to align calculations consistently across the corporation and governance documents to drive a regular process. Our initial work with the EPA Climate Leaders program helped to provide a solid foundation from which to build our next GHG abatement program. Because we consistently advocated for the importance of GHG reductions and low carbon generation, our own internal program had to be sound and credible. The care taken during this early effort has proven to be vital to all of the successes achieved since that time (Fig. 9.1).

---

[4]When demand for electricity goes up, such as during a heat or cold spell, total cumulative emissions go up as well.

Despite the challenge of our first GHG goal, the retirement of a number of fossil steam plants and other business decisions allowed Exelon to make considerable progress. Our initial Climate Leaders goal of 8 % was achieved in 2008 with a 36 % reduction from our 2001 baseline. Our results were verified by a third party and confirmed by the EPA. Inspired by our success and a political atmosphere that was relatively receptive to carbon legislation,[5] we saw an opportunity to become a leader in clean energy while also capturing business advantage.

## 9.3 Integrating Low Carbon into Corporate Strategy

In 2008, through a unique marriage of financial and environmental strategy, *Exelon 2020* was born. Rather than conforming exclusively to evolving GHG accounting standards, *Exelon 2020* was created by looking at electric sector GHG emissions in perspective with business considerations. The corporate strategy department evaluated current and potential regulations and other market drivers shaping business decisions and considered how they also could also contribute to reducing GHG emissions. This broad thinking brought together traditional internal emissions reduction opportunities associated with our own GHG inventory with ways we could help our customers and the affect we could have on the overall electric grid through our nuclear uprates.[6] Our strategy combined state-mandated activities such as customer energy efficiency programs with voluntary ones such as expanding generation capabilities at existing nuclear plants. The combined view assessed what Exelon's overall impact could be, and a key element of the program was the careful consideration given to which projects made the most financial sense.

The final goal was set at 15.7 million metric tons of abatement in a single year by 2020. The company used *Exelon 2020* frequently to highlight the importance of carbon legislation and establish Exelon as a thought leader on this issue. It was embedded in company performance measures, tracked quarterly, and deemed important to all business units and employees. And because the goals for *Exelon 2020* aligned well with key financial performance goals, such as O&M cost control, generation fleet performance, nuclear uprate program completion and utility compliance with RPS and energy efficiency mandates, it would also be good for business (Fig. 9.2).

*Exelon 2020* reached beyond the operational components captured by a traditional GHG inventory, which can limit boundaries to certain lines of business or company assets. The program easily crossed operating company boundaries, and helped to uncover new opportunities for employee engagement and GHG reduction. By thinking beyond the confines of traditional Scope 1 and 2 GHG

---

[5]By "carbon legislation" we mean legislative limits on greenhouse gas emissions.

[6]Nuclear uprates—means increasing the electrical generation capacity of a nuclear power station.

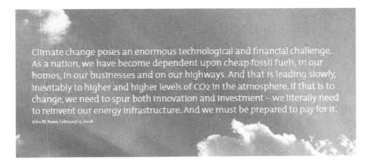

**Fig. 9.2** Excerpt from the original *Exelon 2020*: A Low Carbon Road Map announcing the program in 2008

accounting,[7] *Exelon 2020* became part of the corporate culture and linked it to our broader commitment to environmental excellence. We tied it to our recycling and land resource management programs, and it became a driver of operational efficiency improvements across all levels and locations.

## 9.4 Seeing the Big Picture

Establishing the accounting construct for *Exelon 2020* required creativity. Years before the World Resource Institute introduced its official Scope 3[8] accounting standard, *Exelon 2020* sought to capture exactly that—the impact our actions could having on broader sector level emissions. Exelon utilities ComEd and PECO were providing customers with real energy efficiency savings through their Smart Ideas® customer programs, and more renewable energy through renewable energy certificate (REC) retirements to fulfill state renewable portfolio standards. In addition, our generation company was optimizing its nuclear fleet to further expand low carbon nuclear generation at our plants through uprates—increased capacity at its existing nuclear power plants through equipment optimization and efficiency improvements. Exelon included the associated avoided emissions in its GHG abatement approach in order to highlight the GHG reduction benefit being achieved for customers and the community as a result of these actions.

---

[7]World Resource Institute defines Scope 1 emissions as those direct from a company's operations, while Scope 2 are the indirect emissions associated with the generation of the electricity that company is using.

[8]Scope 3 emissions are those associated with the production, distribution or use of your business product beyond your owned or operated corporate boundary. In this context it refers to the electricity that our customers purchase and use, beyond that which is generated by our generation stations.

Working with outside consultants, Exelon prepared specific protocols documenting the accounting of reductions from customer abatement[9] and nuclear displacement[10] to the Voluntary Carbon Standard (VCS) 2007.1 (version November 18, 2008) and VCS Project Description Template (version November 19, 2007). We established internal management model documents covered under our ISO 14001 Environmental Management System (EMS) that referenced the corporate GHG Inventory Management Plan, quarterly accounting procedures and reporting requirements. Performance was regularly reviewed in quarterly management meetings with executives.

The program reported its progress in GHG emissions abatement or avoidance as a result of actions taken or enabled by Exelon. Each year we worked to increase our level of GHG abatement, contrary to reporting only on our declining emissions. This strategy captured our actual emission reductions (the difference between our current year GHG inventory and that of our 2001 baseline) and added the tonnes of GHG avoided from:

1. Project-based reductions (primarily associated with improved waste management);
2. Customer abatement programs; and
3. Nuclear uprates program.

*Exelon 2020* provided an annual snapshot of company performance. If emissions increased in a given year (due to increased electricity demand or weather, for example), reductions from our Scope 1 and 2 inventory might decrease from the prior year. Similarly RECs[11] purchased for RPS[12] obligations are tied to a specific year and are not cumulative year over year. In contrast, customer energy efficiency carries forward for the life of the efficiency improvement, but the lifetime of the effort varies from action to action. Nuclear uprates would carry forward for the life of the plant. *Exelon 2020* sought to quantify all of the ways that we were directly involved in driving the transition to clean energy and abating GHG emissions, and how these actions worked together.

Viewed through the traditional lens of GHG inventory accounting, a potential criticism *Exelon 2020* faced related to the concept of "business as usual." It could be argued that many of the actions we took would have happened absent *Exelon 2020*. But because energy is our core business, we continually work to position ourselves for success in current and future markets. To a company that subscribes to a carbon-constrained future, all clean energy decisions are "business as usual." So while we

---

[9]"Customer abatement" means the megawatt-hours (MWh) reduced as a result of our electric utilities' mandated energy efficiency programs and fossil emissions avoided as a result of renewable energy credits that we retired on behalf of our customers.

[10]"Nuclear displacement" means MWh of marginal fossil unit dispatch that is avoided as a result of our increased production at our nuclear plants.

[11]A "REC" is a renewable energy credit equal to one MWh of electricity with the environmental attributes of a specific renewable electric generation source.

[12]Renewable portfolio standards (RPS) refers to state-mandated percentages of renewable generation required in the overall mix of coal, nuclear, hydro, etc.

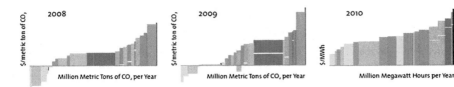

**Fig. 9.3** Excerpt from the 2011 *Exelon 2020* update report: The carbon supply curve provided the carbon price needed to make each of the identified projects economically break-even, with negative carbon prices implying that the associated projects would be economic to undertake even without a price on carbon. In the 2008, 2009 and 2010 reports we published versions of this curve that reflected the then-current market environment, including expectations for commodity prices, load growth, regulatory changes, and costs of new energy demand reduction/supply options

clearly disclosed which elements were mandatory versus voluntary, *Exelon 2020* allowed us to excel in both. Most importantly we continued to focus on solutions driven by competitive markets and that made good financial sense (Fig. 9.3).

Grouping them under the *Exelon 2020* strategy illustrated how important all market forces are to achieving greater GHG reductions:

- When generation plants became inefficient and uneconomical, they were retired and opportunities were sought to replace this generation with cleaner sources.
- Our facilities departments were challenged to reduce energy use in our commercial and support buildings.
- State-mandated programs for energy efficiency and Renewable Portfolio Standards (RPS) were designed to go beyond mere compliance. ComEd's and PECO's Smart Ideas® programs won multiple awards for marketing and customer relations. In addition, our utilities always met their efficiency targets on time and often early.[13]
- Our nuclear uprate program was established as a means to both improve the efficiency of our nuclear plants and expand power capacity to sell.

This strategy demonstrated both what Exelon was doing internally to address emissions and how it was supporting local and state efforts to reduce emissions. It also established GHG accounting for efforts not otherwise being captured.

And most importantly, it inspired our employees. *Exelon 2020* provided a platform for employee engagement around environmental topics. It was a recognizable moniker that connected contests and events with the greater corporate focus on low carbon initiatives. Over the years, there were employee "Green Video" contests, waste reduction challenges, and waste and recycling dumpster dives, all which could be tied to the overall corporate strategy of *Exelon 2020*. In addition, innovative ideas to manage impacts of our supply chain or other operational areas could be more easily communicated when shown to align with or improve *Exelon 2020* performance.

---

[13]Regulated utilities are restricted from exceeding their regulatory targets due to customer rate case agreements that prevent additional investment beyond what has been approved by the local Public Utility Commission.

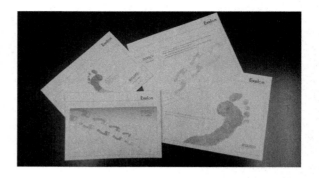

**Fig. 9.4** *Exelon 2020* reports used to relay the corporate progress and position relative to greenhouse gas reduction and potential regulation

Exelon issued an *Exelon 2020* report each year to inform stakeholders and support advocacy for a price on carbon. These reports provided an update on our progress and showed, using a GHG supply abatement curve, exactly how the market was driving our decisions.[14] The reports combined financial and electric market analysis with GHG abatement information to tie together our investor relations and regulatory advocacy activity. *Exelon 2020* provided a long-term view for investors that allowed for opportunistic business decisions driven by market conditions. By communicating our carbon strategy to our stakeholders and making the connection to value, Exelon was able to balance investor and environmental interests and show that action on carbon can be good for business and the environment (Fig. 9.4).

## 9.5 And then a Carbon Price Did Not Happen

Several key events in 2009 and 2010 substantially changed the energy business:

- The USA fell into an economic recession;
- The Waxman-Markey cap and trade bill on carbon emissions was defeated;
- The Production Tax Credit[15] (PTC) was extended and expanded through the American Recovery and Reinvestment Act (ARRA);
- The Climate-gate controversy raised doubt in climate science; and
- Natural gas fracking technology advances caused natural gas prices to fall.

In 2011, with political gridlock in Washington, it was apparent that federal legislation on climate change or carbon emissions would not pass during the regular business planning cycle. While Exelon knew intuitively that low carbon had a long-term business advantage, there was no indication that the federal government would take action in the short term to drive markets in that direction. In fact,

---

[14] In later years, this report became part of Exelon's Corporate Sustainability Report.

[15] The Production Tax Credit (PTC) gave tax advantages to renewable production projects by crediting electric production from these sources regardless of whether that generation was needed on the grid at the time of generation.

> "From Exelon's perspective, climate change is one of the greatest financial, technological and public policy challenges facing our industry and the country."
>
> William A. Von Hoene, Jr., Senior Executive V.P. and Chief Strategy Officer
> Exelon Corporate Website 2010

**Fig. 9.5** Exelon corporate website 2010

disjointed local and state actions were beginning to negatively impact the electric markets where Exelon operates (Fig. 9.5).

Conversations shifted from carbon emissions to energy or monetary savings. The voluntary federal Energy Information Agency (EIA) 1605 (b) greenhouse gas reporting program was halted, and the EPA Climate Leaders program was phased out. However, we continued to work toward our *Exelon 2020* goal, including achieving an interim goal of reducing energy usage in commercial buildings by 25 % from 2001 levels by 2012, which was estimated to be saving $3 million to $4 million in operating costs annually.

Low power prices, depressed load volumes, a patchwork of local and state subsidies and incentives for energy efficiency and the development of distributed generation challenged the *Exelon 2020* plan as it had been designed. It became difficult to gain support for unique energy efficiency projects, and several planned nuclear uprate projects were delayed or canceled. But the program continued to prove its value in many other ways.

## 9.6 The Value of *Exelon 2020*

Beyond direct monetary savings from internal energy efficiency, the *Exelon 2020* program continued to prove its value. Our early efforts in accounting prepared us to comply with the rollout of the EPA's Mandatory GHG reporting requirements under Part 98 of the Clean Air Act. While we incurred some fees in preparing site-specific monitoring plans, many aspects of the program were already in place as a result of our voluntary accounting program.

The *Exelon 2020* program also helped us frame discussions of how our businesses could be more sustainable. It created a strong foundation for evolving a GHG-centric program into one that incorporated other environmental issues that are material to our business. *Exelon 2020* demonstrated the need for cost consciousness and market wisdom in evaluating the three pillars of sustainable energy—clean, reliable, and affordable (Fig. 9.6).

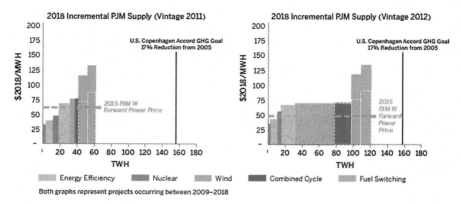

**Fig. 9.6** Excerpt from the Exelon 2012 Corporate Sustainability Report: While low natural gas prices are providing affordable electricity to our customers and lowering GHG emissions, these sustained low prices are also putting financial pressure on energy efficiency measures and deployment of capital into new cleaner, reliable generating capacity. One unintended consequence of these market conditions and policies is their potential to force some nuclear plants to retire prematurely. Low gas prices, coupled with government subsidies for renewable technologies that disregard market demand or market price, are inadvertently putting at risk our nation's progress toward its environmental goals; this calls for a reexamination of current subsidies and incentive policies

## 9.7 Managing the Exelon-Constellation Merger

In March 2012, Exelon completed its merger with Constellation. During the Constellation merger, *Exelon 2020* aligned with the state of Maryland's climate and economic goals and supported the negotiation of commitments for new low carbon generation in the state. As part of the merger agreement, Exelon divested three large coal plants that Constellation had owned, which helped maintain Exelon's low-carbon portfolio. Through the merger, Exelon acquired a limited interest in upstream natural gas developments and a strong customer-facing retail business, which expanded Exelon's competitive energy portfolio, including voluntary REC sales, retail energy efficiency, demand response products, and distributed generation. The companies' complementary businesses and shared focus on environmental performance strengthened our integrated business model and expanded our perspective across the energy value chain.[16]

Constellation Energy had set two GHG reduction goals prior to the merger:

- Enabling customers to avoid 7 million short tons of $CO_2$ on a cumulative basis between 2010 and 2015; and
- Reducing greenhouse gas emission intensity across the generation fleet by 5 % by 2015 compared to 2010 levels.

---

[16]The energy value chain is defined as including upstream fuel development through utility-scale generation, electrical and natural gas distribution, distributed generation, retail sales, and beyond the meter customer services that help manage end-user energy use and cost.

**Fig. 9.7** A 2012 map depicting the combined assets of Exelon and Constellation Energy through the merger

Amidst the change that accompanies any larger merger, Exelon concluded it was important to maintain its existing GHG commitments and worked to integrate Constellation's goals into *Exelon 2020*.

In doing so, Exelon made conservative assumptions when necessary, which it disclosed and explained. The Constellation customer goal was translated into the annual equivalent of abating 1,027,059 metric tons of GHG emissions. Similarly, the goal for the Constellation generation fleet was translated into the annual equivalent of abating 814,261 metric tons $CO_2e$. For scale, one million metric tons of GHG emissions are equivalent to removing approximately 200,000 cars from the road for a year.[17] These annual targets were added to the existing *Exelon 2020* target to establish an updated abatement goal of 17.5 million metric tons $CO_2e$ in a year (Fig. 9.7).

As a result of the integration of the two GHG inventories and associated goals, it also became necessary to ensure that the baselines from which we measured reductions were also maintained. While Exelon continued to report actual third-party verified emissions annually to both The Climate Registry and the Carbon Disclosure Project, within the bounds of *Exelon 2020* separate baselines were maintained at the site level to accurately account for the GHG reductions that Exelon brought about. For the legacy Exelon sites, measurement continued to be against the 2001 baseline; however, for the Constellation sites (partially because full GHG accounting information was not available back to 2001) GHG reductions to be counted toward the *Exelon 2020* program were measured from their third-party verified 2012 baseline. Because *Exelon 2020* captured reductions, maintaining these two baselines through the course of the program was the best way to capture the reductions in GHG emissions brought about by Exelon actions.

---

[17]Per EPA GHG equivalency calculator (4.75 mt $CO_2e$/car/year).

While administratively combining the two programs was complex, clear documentation and disclosure of our intent and our efforts made it possible to do so. In addition, maintaining the well-established program construct from the *Exelon 2020* program greatly helped to build consistency in GHG reporting across the newly merged corporation, integrating assets into the Inventory Management Plan, and supporting accounting development in locations that may have not been included in the Constellation GHG inventory prior to the merger.

Having then achieved the goal within two years of completing the merger and goal update, it could be argued that a more aggressive goal could have been set at the time of the merger. However, the slower and steadier approach taken by Exelon allowed for the corporation to maintain a GHG goal seamlessly during merger integration. In the merger integration process, we began to align the various legacy and new operating companies on a single goal and allowed each company to build on the momentum that they had already created under their established programs.

## 9.8 Achieving *Exelon 2020*

Because Exelon reported progress quarterly, it anticipated achieving the overall *Exelon 2020* goals in 2013. Therefore, as soon as the year ended, Exelon began the third-party verification on not just its Scope 1 and 2 GHG inventories, but all of the elements of the *Exelon 2020* program (Fig. 9.8).

To support the planned Earth Day 2014 announcement, Exelon verified that it had abated 18.1 million metric tons of GHG emissions in 2013 to reach its *Exelon 2020* goal. While different from our initial *Exelon 2020* Low Carbon Road map, the actual journey to achieving the *Exelon 2020* goal was a result of the unforeseen challenges and changes in the electricity markets over the years (Fig. 9.9). How we actually achieved the goal, seven years ahead of schedule, includes the following elements:

- The majority of our abatement (55 %) came from emissions reductions from power plants. This included about 8.2 million metric tons of GHG emissions from the retirement of older, inefficient generating plants that burned fossil fuel. Another 1 million metric tons of reductions came from changes in generation dispatch to cleaner units or less frequent dispatch of fossil peaking plants in 2013. The remaining reductions were achieved through energy efficiency and process improvements within our own operations. Though offset by emission increases in other areas, such as upstream natural gas production, a total 9.8 million metric tons of actual GHG emissions reductions were accounted for in 2013.
- Approximately 1 % of our achievement came from offsets[18] and project-based reductions. This included many of our employee engagement activities, which allowed employees to pursue special projects to contribute to *Exelon 2020*. For example, using EPA's waste management (WARM) model accounting, Exelon's

---

[18]Offsets are credits that document carbon reductions from a specific project.

9  Exelon Corporation: Strategic Greenhouse Gas Management

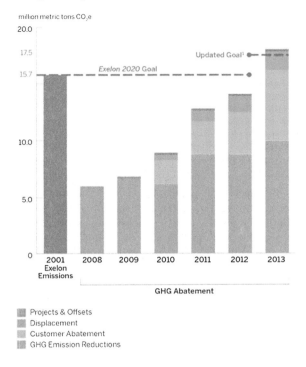

**Fig. 9.8** Excerpt from the 2013 Exelon Corporate Sustainability Report

> "We believe that clean, reliable and affordable energy is key to a more sustainable future. Exelon 2020 has helped us move closer to that vision. It provided us with a clear goal, and we could not have accomplished it without the dedication and effective collaboration of our employees, customers, communities and business partners."
>
> Chris Crane, **Exelon President and Chief Operating Officer**
> **Exelon 2020 achievement press release – April 23, 2014**

**Fig. 9.9** *Exelon 2020* achievement press release—April 23, 2014

supply management department captured GHG benefits from increased office recycling and investment recovery initiatives. Such programs raised employee awareness and spurred site-level energy efficiency efforts.
- Customer programs contributed approximately 35 % of our 2013 performance. Of this, 28 % can be attributed to programs run by our utilities BGE, ComEd, and PECO in accordance with state mandates for customer energy efficiency. Each utility has received the EPA's Energy Star award for performance excellence in efficiency programs. Tailored for residential and business customers, the programs saved approximately 6.3 million megawatt-hours of electricity in 2013.
- The remaining 7 % of the customer abatement progress came from the Constellation business and its retail customer programs. Constellation's retail unit provided energy management products and services to approximately 100,000 commercial and 1 million residential customers, having a tangible impact on greenhouse gas emissions. These programs include energy conservation measures, demand response, customer rooftop solar systems, as well as the sale of 2.4 million voluntary renewable energy credits (RECs).
- Nuclear uprates accounted for 9 % of our GHG abatement, helping Exelon avoid more than 1.6 million metric tons of emissions in 2013. While downward pressure on electric prices resulting from cheap natural gas and market structures that do not recognize the value of reliable clean energy resources hindered our nuclear uprates plan, displaced fossil generation from those implemented still contributed significantly to program performance.

## 9.9 The Corporate Bottom Line for *Exelon 2020*

Although there was never a federal price on carbon enacted during the lifespan of *Exelon 2020*, which would have directly increased the value of our low carbon generation fleet, the program still created value in many ways. Even beyond the direct environmental benefits (a total of 67.8 million metric tons $CO_2e$ abated over the 6 years of the program—equivalent to removing over 14 million cars—18 % of the nation's vehicles—from the road for a year),[19] basic cost cutting associated with internal efficiency improvements (estimated to be over $3 to $4 million annually in purchased electricity alone), and millions in customer monetary savings realized through our customer energy efficiency programs, the *Exelon 2020* program brought together our employees around a common cause and developed an identity for the corporation. Exelon has built a reputation as a leading low carbon energy provider and has an established credibility in this ongoing national discussion that has such critical implications for the energy sector.

Through *Exelon 2020*, we have also gained experience recognizing lowest cost options for sector-wide GHG reductions, as well as the potential for GHG

---

[19]Includes avoided emissions of 18.1 million 2013; 14 million in 2012; 12.9 million in 2011; 8.9 million in 2010; 7.9 million in 2009; and 6 million in 2008.

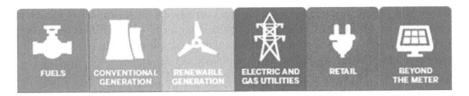

**Fig. 9.10** Exelon's energy value chain perspective

emissions shifting between segments of the energy value chain. Actions within each segment of the energy value chain have implications for the others. With *Exelon 2020*, we were capturing the GHG benefits across these segments, and could begin to see how unintended consequences could arise and create negative GHG impacts as well (Fig. 9.10).

The achievement also highlighted the critical role of the existing nuclear fleet in reducing GHG emissions and has helped us communicate the need to preserve and value no carbon generation. The program's value chain accounting also helped to highlight how disjointed actions on the electric grid (even if well intentioned) can actually lead to GHG increases if not properly integrated both physically and into market structures. *Exelon 2020* has led us to look more critically at our ability to have an impact on grid emissions and how our operations relate to state and US GHG emission goals (Fig. 9.11).

Though still pending implementation, there are many similarities between the design of *Exelon 2020* and the building blocks of EPA's 111(d) regulations as initially proposed in 2014 for existing plants (Clean Power Plan). Intrinsically there is value to having experience with this type of accounting. Reliability and

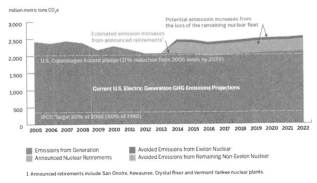

**Fig. 9.11** Excerpt from the 2013 Exelon Corporate Sustainability Report: This graph displays historic and forecasted GHG emissions from the electric power generation industry from 2005 to 2022, based on 2013 US EIA annual energy outlook data along with a representation of GHG emissions increases that could occur if all nuclear generation was retired and replaced by other generation technologies currently available on the grid

affordability of the electric supply are important aspects in the transition to a clean energy future for the USA. The *Exelon 2020* program has prepared us to communicate on these issues and raise awareness that these elements should be properly recognized in policies or implementation plans that may ensue.

## 9.10 Beyond *Exelon 2020*

As we move forward, through *Exelon 2020* and other similar analysis, we recognize that our response to climate change must move beyond emission reduction efforts to include technology, fuel diversity, grid reliability, government policies, and market considerations. We are considering what could happen to grid rates if nuclear generation retires and is replaced by currently available technologies; and we are considering land use impacts, as well as issues of affordability and reliability as we explore new technologies to help transition to lower carbon electric supplies. We will continue to work with stakeholders to focus on the best ways to reduce our emissions, communicate our progress, and track how markets impact emissions for us and the overall electric industry.

Although market conditions continue to change, we maintain focus on carbon emissions as a value driver for our business and recognize the importance of other potential climate change issues on our assets and ability to serve our customers. To this end, we have established a Climate Policy and broadened our climate change response to include actions in three primary focus areas:

1. Continued GHG Emission Reductions from Internal Operations,
2. Contributing to Lower Electric Grid GHG Emissions for our customers,
3. Addressing Infrtructure Resiliency.

Organizing our efforts to advance clean energy around these areas demonstrates that Exelon is not only responding to climate change, but also helping our customers and our country to do the same. *Exelon 2020* was fundamental to the development of our holistic view of GHG emissions reductions and establishing our value chain perspective for driving sustainability. Exelon celebrates the achievement of this goal and encourages others to look broadly for economic opportunities that support social and environmental issues.

> *"We believe that government should play a major role in establishing GHG abatement goals for the nation, but that technology and capital investments to achieve national goals should be left to subsidy-free competitive markets that are best suited to ensuring that least cost compliance is achieved."*
>
> Chris Crane, Exelon President and Chief Operating Officer
> Exelon 2013 Corporate Sustainability Report, Letter from the CEO

**Fig. 9.12** Exelon 2013 Corporate Sustainability Report, letter from the CEO

For us, low carbon electric generation is fundamental to our business. The *Exelon 2020* program was a success and resonated throughout the corporation because it aligned with our business goals and the potential optimization of the financial value of our products and services. Our value chain perspective allowed us to see connections and opportunities that would not have otherwise been as apparent, and with changing times, and shifting social issues and economic conditions, we hope to use this perspective to similarly evolve our programs and electric markets to achieve even further performance.

For others seeking to reduce GHG emissions, we recommend looking at not only your operations, but also to how your products and services connect to markets and your customers. Seek cost-effective solutions that drive value for your product, and if the value of low carbon products or services is not effectively being captured by your existing market structures, seek to change those market structures. Making that connection between GHG reductions and future business value is key if we are to reach the reductions needed to minimize future climate impacts (Fig. 9.12).

# Chapter 10
# Hoosier Energy Rural Electric Cooperative, Inc.: The Rural Cooperative Perspective

## Michalene Reilly

**Abstract** Making the business case for sustainability is far easier for large utilities that have either investors or a significant customer base pressuring particular actions. The business case is more difficult when those drivers are absent, as is the case for small non-investor-owned utilities (non-IOUs). For large non-IOUs, there is still pressure to respond to stakeholders and an extensive customer base, but it is more challenging to promote sustainability inside smaller organizations, such as rural electric cooperatives. Hoosier Energy, located in Bloomington Indiana, has been implementing programs to support its communities since its founding in 1949 but has only recently begun to frame these programs as continuous improvement. It took the development of a plan for new headquarters facilities to help bring all the pieces together into a theme that made sense for Hoosier Energy. The Hoosier Energy story is presented as an example of garnering support for a comprehensive program within a generation and tranmission cooperative .

Making the business case for sustainability can be difficult. It is far easier for large utilities that have either investors or a significant customer base pressuring particular actions. When those drivers are absent, as is the case for smaller non-investor-owned utilities, the business case for sustainability is even more difficult. For large non-IOUs, there is still pressure to respond to engaged stakeholders and an extensive customer base, even if they do not have investors. The question becomes: How do you support sustainability inside small organizations, such as rural electric cooperatives?

M. Reilly (✉)
Environmental Special Projects, Hoosier Energy, Bloomington, USA
e-mail: mreilly@HEPN.com

Hoosier Energy has been a member of the EPRI Sustainability Interest Group work group since its inception. Environmental staff were anxious to capture work already being done at the cooperative. In the heartland, words such as "sustainability" are often replaced with words such as continuous improvement, efficiency and productivity. Pressure from environmental groups that may not take all aspects of sustainability (financial and social) into account often gives the word "sustainability" a black eye. The road to developing an effective sustainability program needs to be slow and measured until a catalyst brings the components together.

The Financial Times Lexicon[1] states that "Business sustainability is often defined as managing the triple bottom line—a process by which companies manage their financial, social and environmental risks, obligations and opportunities." It took the development of a plan for new headquarters facilities to help bring many pieces of sustainability together into a common theme. The Hoosier Energy story is presented as an example of garnering support for a number of initiatives that could be construed as sustainable elements.

## 10.1 Background

There are approximately 840 electric distribution co-ops in the USA distributing electricity to their members and 65 generation and transmission cooperatives, many of which provide generation and transmission assets for their members. Electric cooperatives are as follows:

- Private, independent, and non-profit electric utilities;
- Owned by the customers they serve—their "members";
- Incorporated under the laws of the states in which they operate;
- Established to provide at-cost electric service;
- Governed by a board of directors elected from the membership which sets policies and procedures that are implemented by the co-op's management.[2]

Continuous Improvement/Sustainability can play a very important role in cooperatives that have smaller resource pools. As many larger companies are making the business case for a formal sustainability program, the electric cooperatives can add value from these initiatives better leveraging the resources available at a cooperative business.

Tying together the good works of efficiency, environmental compliance and stewardship, employee engagement, and day-to-day sustainable business practices into a cohesive process can be both challenging and rewarding. Perhaps, the most important question is: If you are a sustainable organization, is it more important that the organization be *classified* as "sustainable" or that the organization is

---

[1] http://lexicon.ft.com/Term?term=business-sustainability.

[2] Information from the National Rural Electric Cooperative Association website: http://www.nreca.coop/about-electric-cooperatives/co-op-facts-figures/.

simply performing the actions that make them effective? The journey of Hoosier Energy Rural Electric Cooperative, Inc. and others who do not have stockholders will be addressed in this chapter. Although investor-owned utilities have many of the same experiences, they are often bolstered by shareholder "encouragement."

## 10.2 The Cooperative Perspective

There are many differences between cooperatives and IOUs. For example, the cooperative business model is nonprofit. The members of the cooperative determine rates through members and the board of directors. Unlike IOUs which often have an equity ratio of 50 % or greater, cooperatives tend to have a much lower equity ratio of around 20 %—which means their facilities are more highly encumbered by loans and other financial transactions. Electric Cooperatives serve an average of 7.4 consumers per mile of line and collect annual revenue of approximately $15,000 per mile of line while investor-owned utilities average 34 customers per mile of line and collect $75,500 per mile (publicly owned utilities, or municipals, average 48 consumers and collect $113,000 per mile). This discrepancy exists despite the fact that cooperatives own and maintain 2.5 million miles, or 42 %, of the nation's electric distribution lines, covering three quarters of the nation's landmass.[3] The rural nature of a cooperative's territory leads to higher costs for the delivery of electricity, while at the same time cooperatives have a constant drive to keep cost of power as low as possible, with the rate payers determining those rates.

Although diverse in their nature, electric cooperatives generally fall into two categories. Distribution cooperatives—those that actually distribute electricity from the substation to individual homes and businesses (although some distribution cooperatives actually own and operate their own substations)—are the first type. These are small local businesses located in their service territory and serve a central role in the community where they operate. Distribution cooperatives may have as few meters as 350 (Alaska) or as many as 250,000 in the largest (Texas), with the average size being 22,800 m.[4] The customer base is predominately residential.[5]

The second category of cooperative is a generation and transmission cooperative (G&T). G&T cooperatives were often formed to serve a group of distribution cooperatives for either the wholesale purchase of electricity, or to develop generation to serve the distribution members (Figs. 10.1 and 10.2).

---

[3]NRECA website on facts and figures http://www.nreca.coop/about-electric-cooperatives/co-op-facts-figures/.

[4]The IOU average size is 540,000 m.

[5]Co-ops have an average of 57 % of their customers as residential versus 43 % commercial and industrial. IOUs serve 36 % residential and 64 % commercial and industrial.

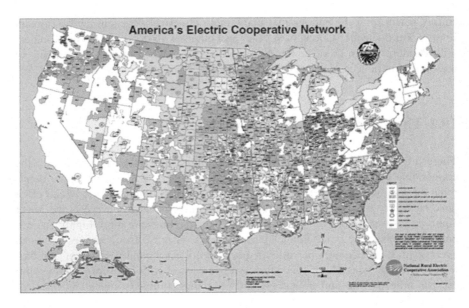

**Fig. 10.1** America's Electric Cooperative Network

**Fig. 10.2** G&T Cooperative Service Areas 2014

The G&Ts in the USA are diverse with 40 of the 65 having their own generation plants and 25 which serve as bulk purchasing agents only. A G&T may serve a single state (or part of a state) or cross several state lines. Its members may serve as few as 110,000 m or in the case of Basin Electric, the largest G&T, 138 members and more than 2.9 million meters.[6] Hoosier Energy is a medium size G&T cooperative that sells power to 18 distribution cooperatives (17 in Indiana and one in Illinois) which serve approximately 300,000 m.

For a cooperative, the word "sustainability" may have different meaning than for an IOU. The rural distribution cooperative with limited resources and few employees has a primary focus to "keep the lights on" at the lowest possible cost. Sustainability may include keeping rates low and serving its customers. The triple bottom line of "social, environmental, and financial" is a constant and driving force for utilities for those "bottom line" reasons. Just because the word "sustainability" has not been linked to the activities or is not part of a report does not mean that many of the accomplishments are not in place.

## 10.3 What Is "Sustainability" to a Cooperative?

Although a cooperative utility may not use the word "sustainability," one look at its annual report will determine whether programs that are being performed can be categorized in that way. Many of the programs discussed below fall under the term "sustainability."

Many cooperatives have historically viewed the word "sustainability" as a term advanced by environmental groups. They believe many environmental organizations view the term as a method to inhibit development. They believe sustainability means less about developing economic growth and more about preventing development in the very communities where distribution cooperatives are located. Therefore, the word "sustainability" carries many negative connotations. And yet, as cited above, many of the same concepts are considered core to being a cooperative. The seven cooperative principles are listed below:

1. **Voluntary and Open Membership**: Cooperatives are voluntary organizations; open to all persons able to use their services and willing to accept the responsibilities of membership, without gender, social, racial, political, or religious discrimination.
2. **Democratic Member Control**: Cooperatives are democratic organizations controlled by their members, who actively participate in setting their policies and making decisions. Men and women serving as elected representatives are accountable to the membership. In primary cooperatives, members have equal voting rights (one member, one vote) and cooperatives at other levels are also organized in a democratic manner.

---

[6]http://www.basinelectric.com/About-Us/.

3. **Member Economic Participation**: Members contribute equitably to, and democratically control, the capital of their cooperative. At least part of that capital is usually the common property of the cooperative. Members usually receive limited compensation, if any, on capital subscribed as a condition of membership. Members allocate surpluses for any or all of the following purposes: developing their cooperative, possibly by setting up reserves, part of which at least would be indivisible; benefiting members in proportion to their transactions with the cooperative; and supporting other activities approved by the membership.
4. **Autonomy and Independence**: Cooperatives are autonomous, self-help organizations controlled by their members. If they enter into agreements with other organizations, including governments, or raise capital from external sources, they do so on terms that ensure democratic control by their members and maintain their cooperative autonomy.
5. **Education, Training, and Information**: Cooperatives provide education and training for their members, elected representatives, managers, and employees, so they can contribute effectively to the development of their cooperatives. They inform the general public—particularly young people and opinion leaders—about the nature and benefits of cooperation.
6. **Cooperation among Cooperatives**: Cooperatives serve their members most effectively and strengthen the cooperative movement by working together through local, national, regional, and international structures.
7. **Concern for Community**: Focusing on member needs, cooperatives work for the sustainable development of their communities through policies approved by their members.[7]

So does sustainability live in a cooperative organization? Currently, cooperatives have many programs that could fall under the label of sustainability. Bringing those various pieces into one overall program can be formidable because not everyone understands the benefits of having a single benchmarking tool for "managing" programs that are unique to different parts of the company. The ultimate goal is to be doing the right things that may end up being part of a broader program related to sustainability.

## 10.4 The Small Non-IOU Triple Bottom Line

The concept of the triple bottom line is not foreign to cooperatives and other non-IOUs. As non-profits, the financial bottom line is to keep rates as low as possible. This not only means keeping costs down, but doing the right things in the right way with the resources available. It means increasing efficiency in the facilities the cooperative owns and providing the materials to make its employees as efficient as possible. It means ensuring environmental compliance to reduce risk while also developing

---

[7]http://www.nreca.coop/about-electric-cooperatives/seven-cooperativeprinciples/.

programs and policies that make sense and support good relationships with regulatory agencies and other stakeholders and taking an active role in how rules are developed within the state. It includes educating the regulators to understand the "costs of compliance" under different rulemaking scenarios. It means educating cooperative members to understand the importance of preserving our natural resources including doing the things that make life easier and better for their members.

The financial "bottom line" is addressed as much by the cooperative principles as anything else cooperatives do. These include a voluntary and open membership, democratic control of the cooperative where elected members are responsible to the membership, members' economic participation which allows for equitable contributions and democratic control of the cooperative's assets, and finally, autonomy and independence which allows for the members to decide on capital improvements. As non-profit organizations, keeping the costs as low as possible while fulfilling their mission to their community at large is central to what cooperatives do.

In regard to the social "bottom line," in addition to electric service, electric cooperatives and other small utilities become deeply involved in their communities by promoting development and revitalization projects, small businesses, job creation, improvement of water and sewer systems, and assistance in delivery of health care and educational services. In fact, within the seven cooperative basic principles, a core cooperative value is "concern for community."

The environmental "bottom line" is addressed by more than compliance. Cooperatives and other non-IOU's throughout the USA work diligently on compliance issues, but the unique member-driven nature of cooperatives results in community action that has added value for both members and the environment.

## 10.5 Hoosier Energy Story

Hoosier Energy is a generation and transmission cooperative that came into existence in 1949 when a dozen distribution cooperative members met in Wilhelm's Café in Rushville, Indiana, and signed articles of incorporation to form Hoosier Energy as the purchasing agent for bulk power for the eight Indiana and one Ohio distribution cooperative. The single Ohio cooperative was forced to withdraw from the agreement due to complications of interstate commerce regulations. A manager was chosen for the G&T and Hoosier Energy began in a small office in Bloomington, Indiana. Over the next 65 years, the power supplier phased in the operation of eight power plants, including the 250-MW Ratts Generating Station (1970) in Pike County, Indiana, and the 1070-MW Merom Generating Station (1982 and 1983) in Sullivan County, Indiana.

The triple bottom line is reflected in the choice of generation owned by Hoosier Energy. In the heart of coal-rich Indiana, Hoosier Energy built its largest asset during the time of the Fuel Use Act that required the utilization of coal rather than gas. As a result, Hoosier Energy's generation portfolio was 100 % coal for many years. Today, Hoosier Energy's generation resources include 50 % ownership of the 630-MW Holland Energy combined-cycle plant in east-central Illinois (purchased in 2009) that uses natural gas as its fuel source as well as two natural gas peaking plants: the 174-MW Worthington Generating Station (2003) and 66 % of the 258-MW Lawrence County Generating Station (2005), both located in Indiana. The 15 MW landfill gas renewable energy facility located in Pontiac IL, a 16 MW Ocharard Hills landfill project in Davis Junction, IL, a 3.2 MW Clark-Floyd landfill project in Borden, In and the 13 MW Osprey Point coal bed methane renewable energy station round out the current generation portfolio. Hoosier has remained heavily invested in delivering coal-fired electricity (because of the realities of economic dispatch and issues with gas delivery), the percentage of assets that are coal fired continues to diminish (Fig. 10.3).

Although it is more difficult for a smaller utility to have a diverse portfolio because of the limited need for generation dependent on the customer base, through planning and careful purchase and development, risk is reduced to members through developing a portfolio that recognizes changes in demand as well as national environmental priorities. In addition, as a small utility that may mean developing resources that are not appealing to a larger utility and keeping the cost to the consumer lower while taking advantage of underutilized resources.

Hoosier Energy has also made significant progress working to increase renewable energy resources in its generation portfolio, including developing two landfill methane generating stations and a coal-bed methane facility and also entering into long-term hydropower and wind energy purchase power agreements. Despite

**Fig. 10.3** The lobby focal wall at Hoosier Energy's new headquarters facility is a tribute to its assets. Generation from fossil fuels and renewables, transmission, as well as water chemistry are highlighted. Three video screens add dynamic content and curently show the Merom station, the limestone pile at Merom, and a wind turbine turning in the wind

being in a state where there is no renewable portfolio standard, Hoosier Energy adopted a Renewable Energy Policy in 2006 that established goals to increase the share of renewable energy in its power supply portfolio. The G&T recognized the advantages of exploring integration of renewables into the power supply portfolio, and in 2006 adopted a voluntary renewables goal. The G&T has moved aggressively to expand renewables in its generation portfolio without being subject to any regulatory requirements or mandates. The commitment to renewable energy has been strengthened with plans for utility-scale solar projects and a voluntary board policy to increase the renewable portion of the portfolio to 10 % by 2025 after realization of the first target.[8]

In 2013, Hoosier Energy completed and began operating the 13-MW Osprey Point Renewable Energy Station in Sullivan County adjacent to the Merom Generating Station. The unique coal-bed methane generation project uses methane from underground coal seams to produce electricity. Osprey Point Station is the first North American coal-bed methane project to deliver electricity directly to the power grid.

The plant was developed based on studies that showed that methane from coal seams was leaking into the atmosphere on the Hoosier Energy property adjacent to the Merom Generating Station. The plant utilizes methane to produce power, the same methane that was being wasted by venting into the atmosphere.

Another part of this renewable energy program includes the implementation of a small-scale solar and wind renewable energy program with the goal of gaining real-world information on costs and operating data for small-scale residential wind and solar installations. This project is designed to share information with distribution cooperative members and the general public in order to start an effective and educated conversation about renewable energy. Hoosier Energy's website[9] has information on each of the sites that was developed showing both the costs of the installation along with the current and historic operating times and the amount of energy generated in order to demonstrate to consumers the actual costs and impacts of these installations. In addition, Hoosier Energy provided grants to members to develop small wind and solar stations at their facilities to demonstrate to members all aspects, positive and negative, of residential renewables. Solar and wind developers often make sales to consumers based on the idea of energy independence and making financial gains from the home installations. Being able to explain the economic realities to members and helping them find the residential solutions that are both economically viable and environmentally beneficial is critical. Indiana has had significant discussions on the issue of net metering. It is difficult for members to understand why they cannot just earn the same amount of money/credit on their bills for renewables as the standard rates for electricity delivered to their homes. Understanding the infrastructure costs and the realities of intermittent electricity provided by wind and solar in a state like Indiana is critical for members before they make large investments.

---

[8]Hoosier Energy 65th Anniversary/2014, 65 YEARS OF COOPERATIVE PARTNERSHIP *An Illustrated History of Hoosier Energy Rural Electric Cooperative*, Published by Hoosier Energy.

[9]http://www.hepn.com/renewablespilot.asp.

When prices for utility-scale solar projects—facilities with at least one megawatt of generating capacity—dropped 60 % from 2011 to 2014, Hoosier Energy began exploring that resource option as well. As an intermittent resource, solar potentially offers small increments of power when it may be needed most, such as on hot summer afternoons. In recognition of this potential value and as a further commitment to renewable energy, the Hoosier Energy Board of Directors approved a 10-MW solar program in July 2014 to serve the 18 member cooperatives. 10 one-megawatt solar facilities are planned for installation at sites across member service territories.

As of 2014, Hoosier Energy's 1250 MW resource portfolio has transitioned from 100 % coal to a 2100 MW portfolio that includes 64 % coal, 33 % natural gas, and 3 % renewables. With the closure of the Frank E. Ratts Station in 2015, the coal mix drops to 59 % coal, 38 % gas, and 3 % traditional renewables. Not only does this change in fuel mixture recognize the changing needs of the planet, but it also recognizes the social, political, and financial well-being of the company. Hoosier Energy and its members continue to support a diverse power supply mix that balances reliability, cost, and environmental concerns—an "all-of-the-above" approach to meeting member power supply needs. As President and Chief Executive Officer Steve Smith pointed out at the 2014 annual meeting, "We have no perfect energy choices. All are flawed. All are needed."

Renewable energy is often very narrowly defined. When Hoosier Energy decided to invest in renewable energy, its goal was to find projects that were both cost effective to its consumers and environmentally beneficial. Some projects do not meet the definition of "renewables" in all states. Hoosier Energy has invested in several landfill methane facilities. There are significant reasons to develop these sites. At landfills without a generation station, methane is destroyed through the use of flares, thus ignoring the waste of the thermal resource that is being vented to the atmosphere. By adding turbine generators, the methane generated is not only destroyed, but energy is produced to replace other forms of electric generation. An added benefit is that the fuel supply is constant—not dependent on the sun shining or the wind blowing. There is a benefit on all levels from this type of generation.

Along the way toward more diversity in the generation fleet, Hoosier Energy developed other activities that, taken together, form the heart of sustainability. The next section talks briefly about the development of these different activities and emphasizes how an organization can bring all of these individual activities together (Fig. 10.4).

**Fig. 10.4** Hoosier Energy receives purchased power from Dayton Hydro, the Rail Splitter wind farm and its own Osprey Point coal-bed methane generating station

## 10.6 The Triple Bottom Line "Timeline"

With respect to the social and financial "bottom line(s)," in addition to providing electric service, Hoosier Energy and its members are deeply involved in their communities by promoting development and revitalization projects, small businesses, job creation, improvement of water and sewer systems, and assistance in the delivery of health care and education services for both the community and employees. Within the 7 cooperative basic principles, a core value is "concern for community." The environmental bottom line begins with exceeding compliance requirements and culminates with an environmental education program developed internally. Below is a summary of some of the major initiatives that exemplify those efforts.

In 1972, Hoosier Energy developed the first Indiana Festival Guide to promote Indiana Tourism. Over the years, Hoosier Energy has continued to team up with the Indiana Department of Tourism on the guide and has financially supported it ever since.

Ensuring a pool of well-trained electrical employees is critical to the G&T and its members. Among the most innovative employee programs in place is the Hoosier Energy Apprenticeship Training & Safety (HEATS) program. The HEATS program was started in 1975 with the goal of providing formal apprenticeship training. The HEATS program is administered by a committee made up of representatives of members, management, and labor. The four-year program involves 144 hours of coursework and 8000 hours of on-the-job training. Safety and training courses are part of the program for apprentices, while skill improvement training is for experienced line personnel. Almost 600 Hoosier Energy and member employees have completed the program since 1975, with additional power plant personnel completing a special HEATS curriculum for power plant apprentices. The program is open to member employees and employees at the G&T.

Hoosier Energy has a long history of solid economic development achievements since the inception of the program in 1985. The economic development department has focused on existing industry retention and expansion, new industry recruitment, industrial site development, and local economic development organization support. The Hoosier Energy Economic Development program is ranked as one of the Midwest's leading economic development organizations. In 2012 and 2013, through the efforts of the Hoosier Energy Economic Development department, more than 1200 new jobs were created in rural Indiana and Illinois with $500 million in new capital investment made in member territories.

Members and their communities benefit from the availability of good paying jobs, tax revenue, and better quality of life because of the jobs that are brought in through economic development programs. Rural communities, many suffering from urban flight, are able to bring good paying jobs closer to home, resulting in stronger, more sustainable rural communities. One little discussed benefit of reversing urban flight is the reduction in emissions from vehicles as many rural residents who commute long distances for jobs that can support their families, with economic development projects, can get good jobs closer to home.

**Fig. 10.5** The Environmental Education center is located at the boat launch to Hoosier Energy's Turtle Creek Reservoir (*left*). The wind turbine and solar panels power the building and put excess electricity on the grid. *Right* Hoosier Energy educator with students making "solar jars."

Being proactive not only allows for increased efficiencies and environmental protection, but when well thought out, can have additional benefits. EPA proposed a rule that could potentially require substations to have secondary containment structures installed. The G&T recognized both financial and environmental benefits to the program. Substations were classified by risk—which could result in release to waters of the state and installed secondary containment structures. Any new or modified substation includes secondary containment. Hoosier Energy has continued to install physical secondary containment at substations as a best practice. An additional benefit was realized for workers who were standing on sturdy steel grating instead of the rock that could easily shift as they worked. Foot and ankle injuries in substations were reduced substantially.

As a rural cooperative, Hoosier Energy noted different opportunities for hands-on resources for science education in its communities. Hoosier Energy developed an Environmental Education Center in 1994 which is located near the boat launch to the Turtle Creek Reservoir (Fig. 10.5). Hoosier Energy built the reservoir as a cooling pond for the Merom Generating Station. The center has been offered free to groups as a meeting location as well as to school groups wanting hands-on science activities. Hoosier Energy environmental staff taught classes for many years until a full-time educator was hired. Fostered by the enthusiasm of management and staff alike, the program was bound to grow and the growing number of materials developed into an online lending library. Named after the education program's mascot, Freshwater Fred,[10] the lending library used materials gathered for the education center to loan out to educators throughout Indiana. Between the years 2000 and 2014, the library has loaned out

---

[10]During the first year of operation of the education center, any student who came into the center could submit their idea for a name for the "environmental turtle" mascot. The winning name, Freshwater Fred, was chosen by three judges from Indiana's Departments of education, natural resources, and environmental management

more than 20,000 pieces of video, software, and curriculum to over 1800 educators in Indiana and surrounding states. Home school educators who have supplemented their children's education with the materials have donated their libraries to Hoosier Energy when their children have graduated to allow more materials to be available. Today, the environmental educator visits schools as well as utilizing the environmental education center and will be utilizing the new headquarters as another location for bringing in school groups. To complement this opportunity, Hoosier Energy is planting an arboretum on the facility for tree identification and is also providing a butterfly garden and areas for identifying native plants. Hoosier is ensuring strong science and balanced policy information is getting to rural students, some of who will work for Hoosier Energy in the future, is an added benefit.

Over the years, Hoosier Energy had noted significant erosion on the shores of the Turtle Creek Reservoir (cooling pond for the Merom Generating Station). Also noted was significant siltation from runoff upstream of the reservoir. Therefore in 1998, Hoosier Energy held a town meeting to determine whether residents and other affected stakeholders within the Turtle Creek watershed would like to form a partnership to improve water quality. The result of the meeting was the formation of the Partnership for Turtle Creek, a watershed group that met for 10 years before merging with a larger watershed group. Hoosier Energy has been a prime player in these groups by providing significant cost share funding in both cash and in-kind donations, as well as bringing expertise to the committee to assist in its work. In 2005, the Partnership for Turtle Creek received the national NRCS Excellence in Conservation award, with specific emphasis on the public/private partnerships that had developed in the group (Fig. 10.6).

**Fig. 10.6** Students looking for macroinvertebrates during a sampling session on the floating dock for the environmental education center, while Freshwater Fred, Hoosier Energy's education mascot, assists a young man generating electricity on the energy bike on Earth Day

## 10.6.1 Additional Involvement in the Community

Commitment to the community also comes in the form of participation in programs to assist the community. Some examples include the YMCA Corporate Challenge and Hoosier Energy's Outrun Cancer and the annual Women's Build with Habitat for Humanity. Hoosier Energy also supports employee-driven activities which include holiday collections for the local food banks and gifts for the underprivileged. These activities not only improve the "quality of life" within the company, but also benefit the community and the activities that employees support personally (Fig. 10.7).

A myriad of other programs are offered through Hoosier Energy's member distribution cooperatives as well as other cooperatives throughout the country. Distribution cooperatives each year sponsor 1500 students to attend the Electric Cooperative Washington Youth Tour, an opportunity for students to meet elected officials and visit monuments and museums. Many cooperatives offer college scholarships for students in their territories. Operation Roundup® is a program used by many cooperatives to fund grants within their community by offering members the opportunity to round up their bills to the nearest dollar each month, donating those funds to worthy community causes. Other programs include Touchstone Energy Camp, an Indiana program that allows sixth graders to learn about how the electric system works as well as enjoy normal camp activities. Community involvement is central to what cooperatives are all about.

Since 2008, other internal programs stressing efficiency have been developed. Hoosier Energy developed an internal efficiency team. The team is called Cost Management and Efficiency Team (CMET), and its charge is to promote and recognize efficiency across the company. These efficiency measures can run the gamut from lower cost office supplies to developing a project management system with the assistance of Indiana University's Kelley School of Business which ensures full evaluation of projects from the environmental, cost effectiveness, and life cycle standpoints.

**Fig. 10.7** Habitat for Humanity women's build 2014

**Fig. 10.8** Saving through effective, efficient decisions

An internal team with a senior staff "champion" guides the CMET team to promote, evaluate, and direct implementation of different initiatives that will increase efficiency throughout the company. Among those initiatives is a program called Savings through Effective Efficient Decisions (SEED). Every employee is encouraged to bring ideas to the team for evaluation. These ideas are not meant to be limited to large power plant-related savings or efficiencies. Evaluation is done on every level of cost within the company where an employee recommends a change in procedure/supplier/other activities. The team evaluates each submittal and assigns those that are recommended to the person or persons responsible for implementing the improvement. The employee who submits the improvement is recognized in Hoosier Energy publications and an annual efficiency video. Over $32 million in savings have been realized in the first seven years of this program (Fig. 10.8).

Besides SEED, another initiative that has resulted from the CMET team in 2014 has been examination of a set of techniques and tools for process improvement. This included looking at various process improvement programs on the market to help take existing efforts from good to great.

Hoosier Energy also assists its members with an extensive energy efficiency and demand side management program. In 2009, enhanced energy efficiency programs were launched to help consumers more effectively manage energy usage and reduce the impact of increasing power costs on consumer rates. With a goal to reduce system peak demand and total energy sales by 5 % from forecasted levels by 2018, Hoosier Energy offered member cooperatives the option of participating in 11 different programs in 2014. The programs change annually to meet member system needs. This is a completely voluntary program as the Indiana requirements have always given an exemption for consumer-owned cooperatives from required energy efficiency programs. However, the Indiana Utility Regulatory Commission (IURC) has recognized Hoosier Energy's program as a model for others in the state.[11]

Members of cooperatives are better able to "control" their personal financial destiny with programs that lower energy costs and place more disposable income in their pockets. But it should also be noted that sometimes consumers choose

---

[11]Significant information has been included from Hoosier Energy Resources. They include the Hoosier Energy newsletter "Energy Lines," the Hoosier Energy website (http://www.hepn.com), and Hoosier Energy 65th Anniversary book, 2014.

**Fig. 10.9** An energy efficiency specialist checks for air leaks and effectiveness of a window seal. Members use mobile energy wall to demonstrate energy efficiency potential to consumers

**Fig. 10.10** Touchstone energy home program

quality-of-life improvements in place of lower costs.[12] In addition, the reduction in energy usage benefits the environment with reduced emissions both from power plants (Fig. 10.9).

An additional program that Hoosier Energy promotes is the Touchstone Energy Home program. A Touchstone Energy® Home is designed to be energy efficient and economical, as well as comfortable. The Touchstone Energy® Home Program represents a national standard based on the 2009 International Energy Conservation Code (IECC). Refer to the code document for additional guidance and clarification. Standards for Touchstone Energy Homes must be met or exceeded to become certified (Fig. 10.10).[13]

---

[12]Hoosier Energy studies have shown that improved heating systems and insulation of homes often results in members being able to heat and cool their homes to more moderate temperatures rather than reducing their energy consumption. In essence, they are choosing to spend the same amount for energy while being able to live more comfortably.

[13]Touchstone Energy Homes information can be found at: http://www.touchstoneenergy.com/homeprogram.

Internal to Hoosier Energy and most cooperatives, there are a myriad of programs that assist employees (such as tuition reimbursement and employee EAP and wellness programs), to assistance within the community for everything from sponsoring local sports teams to funding new sound systems for a local rural school.

## *10.6.2 Internal and External Stakeholders*

The G&T reached out to Indiana University's Kelley School of Business to help design an internal leadership development program. The program provides a cross-functional framework for staff from different business areas to learn management practices and work together to solve corporate problems. After determining the benefit of the program to employees, the program has been offered to member cooperative personnel. Employees have also been given the opportunity to attend the Bell Leadership Academy.

Hoosier Energy partners with Purdue University to donate property and time to the development of hybrid poplars with a large uptake for water. In addition, Hoosier Energy planted 48 acres at the Merom property with native grasses as a sanctuary for wild turkey and deer.

Employee training and development programs have grown in both scope and depth throughout the 2000s, helping employees work more effectively cross-functionally and creating greater workflow efficiencies. The G&T's range of offerings included courses in the fundamentals of leadership, project management, and advanced leadership. By 2014, more than 200 employees had completed the fundamentals curriculum since it began in 2006, and 25 had graduated from the executive leadership program that began in 2011. Since 2006, more than 40 % of the G&T's employees have completed project management training and 85 have participated in the power supplier's tuition reimbursement program. Recently, seventeen Member system employees graduated from the Cooperative Advanced Leadership Development program which is provided through a partnership with the Indiana University Kelley School of Business.

These are only a sampling of the programs Hoosier Energy participates in or has developed over the years. The unique member-driven nature of cooperatives results in community action that has added value for our members and the environment. As part of the energy efficiency programs mentioned above, the G&T applied for and received federal grants to insulate homes in rural areas. Since the proportion of the population that is below the persistent poverty level is higher in many rural areas, this allowed members who could not afford matching grants to reduce their energy usage. Participation and sponsorship in local environmental programs such as earth day and Arbor Day are part of many local cooperative's annual events. Hoosier Energy participates in county-wide river rafting trips that teach eighth grade students about sustainability and environmental protection of the rivers as well as skills such as geolocation. The information is structured so that it directly translates to their activities at home including how what they do in their own yards affects that watershed and why reusing and recycling is so important.

All of these programs could easily fit under the term sustainability. Still, when the word "sustainability" is used, the message often takes on a strong environmental overtone. Sometimes, a single event can bring all those pieces together.

## 10.7 A Catalyst for Sustainability

In 2011, it was determined that existing facilities housing Hoosier Energy's main office and its power delivery and operations teams were no longer adequate to meet the needs of Hoosier Energy and its members. Over 60 sites were examined with the initial thought of keeping facilities together. Given the necessary requirements for all segments of the business, it was determined that two separate locations would better serve the cooperatives.

In 2012, Hoosier Energy began construction of the new power delivery operations center. It was a project requirement that the facility be energy efficient, and substantial employee input was sought on many aspects of construction. For example, when the design for a secondary containment system was presented to the power delivery and environmental personnel, it was determined immediately that the original design would not be user friendly. Personnel came up with a new design that would not only work for the non-energized equipment to insure containment in the event of a release, but would also insure that employees could access equipment safely and efficiently to insure loading and unloading would not be hampered by the containment structure.

Lessons learned during design and construction of the facility were then considered in the design and construction of a new headquarters facility. Hoosier Energy moved into a new headquarters building in December 2014. The plan was to certify the building as a LEED certified facility (Leadership in Energy and Environmental Design). As Hoosier Energy considered the requirements for LEED Gold, it became clear that achieving Gold was possible with the design and implementation of the building as planned. LEED Gold was achieved in April 2015. With preparation for LEED certification, programs that had started internally over the years, such as a recycling program, could be improved and integrated into a formal facilities process.

The 83,000-square-foot building located on Cooperative Way sits on a wooded, 12-acre tract in Bloomington, Indiana. It was designed by local architects with extensive input from employees to achieve the dual purposes of improving workforce productivity as well as providing a model for energy efficiency and environmental stewardship. The building is 48 % more energy efficient than a structure built to conventional standards with a cornerstone for efficiency—a geothermal heating and cooling system that consists of 120 wells drilled 300 feet deep adjacent to the new headquarters. The interior features recycled materials including reclaimed wood from trees removed during construction. The building will use about half the energy per square foot of current facilities by utilizing light-emitting diode (LED) lighting, geothermal heating and cooling, and other energy efficiency features. All storm water from the parking lots is captured in retention ponds that use natural plants and diffusers to not only clean the water of pollutants, but also reduce the velocity of storm water leaving the site to ensure no erosion is occurring off-site. Much of the parking lots utilize permeable paving to reduce storm water volume. The site has six electric car charging stations and bike racks and lockers for employees. In recognition of Hoosier Energy's wellness program, not only is there a workout room for employees and family members 16 and older (and retirees), but Hoosier Energy has provided assistance to teach employees how to use equipment properly. A walking path was developed that allows employees to walk down to the nearby city lake and bike path. The entire facility uses hypoallergenic soaps and cleaners in recognition of employees with allergies and asthma. The well-being of employees was considered as well as the overall environmental and energy impacts (Fig. 10.11).

The new headquarters also speaks about how the G&T began. Displays in the hallways of the headquarters recall the history of the rural electric movement, including a lunch room named for the restaurant, Wilhem's Café, where southern Indiana cooperatives made the decision to form Hoosier Energy to reduce costs for power to each of the distribution cooperatives (Fig. 10.12). This accomplishment, through cooperation among cooperatives, is described in Hoosier Energy's 65th Corporate History book.

**Fig. 10.11** *Clockwise from the top-left* front of building with the bronze statue showing cooperative roots. Electric car charging station. Board room that can serve multi-purposes including dividing for education center classes and training sessions. A look at the equipment room for the geothermal system. One of two rain gardens capturing all runoff from parking lots. Note the porous pavement in parking stalls

**Fig. 10.12** Remembering our roots includes displays showing REA lineman's truck and a history wall with artifacts from early cooperative history in Indiana

Efficiency comes in many sizes. Great consideration was given to efficiency of the building for the workforce. Departments were set up to ensure effective collaboration of those who work together, and workstations were also configured to promote collaboration. "Huddle rooms" are on each floor to allow for impromptu meetings, and collaboration rooms with coffee stations were developed to allow for those short conversations and chance encounters with fellow employees that allow for the effective exchange of information. A sound masking system is used to ensure the ability to work without constant noise interruption.

Glass office walls were constructed to supplement lighting efficiency and provide the opportunity to see fellow employees going past and invite them into discuss business issues. The result is increased efficiency and collaboration.

## 10.8 What Does A Culture of Execution Look like?

The short answer is that a sustainable workforce is a team, or more precisely, a team of teams. Good ideas come from all directions. In fact, all ideas can be discussed, examined, and then discarded if they cannot be improved upon while learning the lessons they teach. Employees are encouraged to bring ideas to increase efficiency and promote innovation. Employees are recognized in publications for their efforts inside and outside of the company. Efficiency efforts are recognized in both written and video presentations. The video presentation is shown throughout the company and shared annually with the board of directors. Cross-functional teams address all aspects of the business and analyze risk, develop IT solutions, and recommend operational improvements across the company.

In addition, teams work with members to ensure that all member services truly meet the needs of each individual member cooperative.

Several programs are employee driven such as the recycling programs and annual Christmas charity collections. Employees established the program and are constantly contributing to its success. Partnering with the local county recycling district to collect recyclables and the local food bank for holiday collections are two examples. Contributing to local programs helps the community to be more self-sufficient. Employees set up and administer electronics recycling as well as an office battery recycling program. Employees are empowered by what is called internally a "culture of execution" to recommend, develop, and implement programs that complement and expand the ideals embodied in the Triple Bottom Line.

The advantage for a cooperative in developing a sustainable culture is that, although the jargon of the sustainability movement is not always utilized, many of the activities have always existed. Finding ways to formalize our efforts, build on them through experience of others, and report those efforts to stakeholders is a next logical step.

## 10.9 Clarity of Purpose

The complexity of the utility industry today could lead to less clarity about the business model of tomorrow, including what it means to be a sustainable utility. A cooperative's purpose remains abundantly clear: to provide affordable and reliable energy to members. For Hoosier Energy, the future is communicated in the cooperative business model advantage. "We always ask the question: 'Is it good for the members?,'" says President and Chief Executive Officer Steve Smith. "That brings clarity of purpose. We have a strong position today, and with our members, a solid platform for the future."

# Chapter 11
# Los Angeles Department of Water and Power: Energy Efficiency for Our City and Our Customers

Maria Sison-Roces and David Jacot

**Abstract** For the Los Angeles Department of Water and Power (LADWP), sustainability has meant developing both a portfolio of energy efficiency programs for customers and an energy efficiency management program for LADWP facilities driven by its sustainability principles. In 2012, as energy rates were rising, the challenge was to help the City of Los Angeles' very diverse customer base control monthly energy costs through programs that would ensure a cost-effective use of customer funds. The solution? A customer-focused portfolio of energy efficiency programs targeting key opportunity areas. For LADWP's facilities, the challenge was accelerating the internal implementation of energy efficiency measures to reduce facility operating costs and lead the way in energy efficiency. The solution was the development of a facilities energy management program that included participation of a cross section of departments within the organization. LADWP realized considerable synergies in responding to these two challenges and now possesses greater capacity to meet future sustainability goals.

## 11.1 Introduction

The power industry is undergoing a complete transformation, and in Los Angeles, LADWP is at the epicenter. New laws change how we provide service and what our power system needs to be. LADWP will replace nearly 72 % of its existing

M. Sison-Roces (✉)
Corporate Sustainability Programs, LADWP, 111 North Hope Street, #1019, Los Angeles, California, USA
e-mail: Maria.Sison-Roces@ladwp.com

D. Jacot
Efficiency Solutions, LADWP, 111 North Hope Street, #1057, Los Angeles, California, USA
e-mail: David.Jacot@ladwp.com

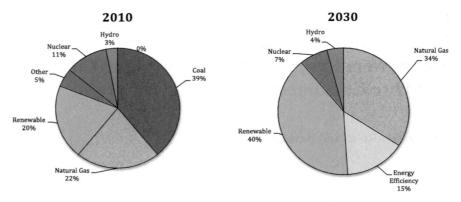

**Fig. 11.1** LADWP Energy Supplies in 2010 and in the Future. *Data source* 2014 Power Integrated Resource Plan (IRP)

electricity supply, resulting in a portfolio with 40 % renewable energy and zero generation from coal, all while removing ocean water cooling from its coastal power plants. For reference, in 2010, approximately 39 % of LADWP's power came from coal-fired plants and 3 out of 4 of LADWP's natural gas power plants used ocean water (Fig. 11.1).

To meet its goal, LADWP will replace 100 % of coal use through renewable energy, greater energy efficiency, and natural gas. Energy efficiency will, for the first time, be relied upon to take a larger role in the energy supply portfolio, increasing to 15 % of supply by 2020. Energy efficiency now plays a vital role in meeting future demand and is the cornerstone of helping customers to manage their energy usage and costs.

As it is not an investor-owned utility, but rather a municipally owned utility, LADWP does not have shareholders to answer to. Instead, LADWP must answer to a wide body of stakeholders including the citizen Board of Water and Power Commissioners, customers, environmental groups, business groups, community groups, Neighborhood Councils,[1] the Los Angeles City Council, the Mayor's Office, and others. With a service territory synonymous with the boundaries of the city, LADWP's operations often become highly politicized.

In 2011, when operating costs necessitated an increase in electricity rates, a well-planned strategy was required. A strategy that would inform stakeholders provides opportunities for their input and enables incorporation of their priorities into the rate increase structure. Stakeholders clearly and consistently indicated that they wanted increased energy efficiency programs to enable customers to manage electricity costs by reducing their energy consumption. They also wanted those programs to generate local economic development.

---

[1] In 1999, the City of Los Angeles adopted a City Charter that established Neighborhood Councils, locally elected bodies empowered to advocate for change in their communities. There are currently 96 Neighborhood Councils in Los Angeles made up of Board Members representing residents, businesses, community interest groups, and property owners.

## 11.2 Background

LADWP is the largest municipally owned utility in the nation, providing reliable energy and water services to 4.2 million residents and 450,000 businesses in the City of Los Angeles as well as 5,000 customers in Owens Valley. LADWP is a proprietary department of the City of Los Angeles, guided by a five-member Board of Water and Power Commissioners who are appointed by the Mayor and confirmed by the City Council. The Commissioners serve five-year terms, are traditionally selected from among prominent business, professional, and civic leaders, and serve on a voluntary basis. The management and operation of LADWP is under the direction of the General Manager, who is appointed by the Mayor and confirmed by the City Council.

### *11.2.1 Customer Base*

The City of Los Angeles is made up of a diverse customer base.
Residential

- Housing—62 % of Los Angeles residents are renters, with as much as 95 % of residents renting in some areas and as little as 8.2 % renting in other parts of the city. 54.4 % of housing units are in multi-unit buildings, which consist of 2 or more units.
- Culture and Language—Los Angeles is a city of diverse cultural backgrounds. 39 % of residents are foreign born, and over 90 languages are spoken in the homes of school age children.
- Education—Adults with a high school education or higher make up an average of 74 % of the population and those with a bachelor's degree or higher make up 31 %. However, the range of education level varies widely from neighborhood to neighborhood, with the percentage of residents with high school diplomas ranging from 25 to 95 % in various neighborhoods and bachelor's degrees ranging from a low of 2 % to a high of 37 %.
- Economic—Median household income for the period of 2008–2012 was $49,745 for the city as a whole and ranged from a low of $22,000 to a high of $207,000 by neighborhood.

Commercial, Industrial, and Institutional

- Commercial—The commercial sector includes the full range of business interests, from retail and restaurants to office buildings. Common areas and commonly metered units of multifamily buildings are also included in this sector.
- Industrial—Los Angeles has a wide range of industrial uses including many with unique characteristics such as manufacturing, movie and television studios, mining, and transportation.

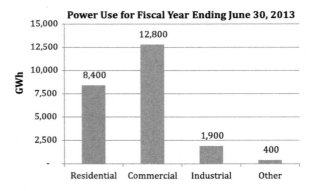

Fig. 11.2 Annual Power Use by Customer Sector

- Institutional—Institutional customers include hospitals, government buildings, colleges and universities, and LADWP's largest customer, the Los Angeles Unified School District.

Business and industry consume about 65 % of the electricity in Los Angeles, but residents constitute the largest number of customers (Fig. 11.2). Average residential energy use is about 500 kWh per month (2014 Integrated Resource Plan). Total annual energy delivery within LADWP territory is 23,500 GWh to 1.4 million electric service connections.

## 11.3 Customer Program

The current energy efficiency program was an outgrowth of the rate increase process that LADWP embarked on in 2011. The atmosphere at the beginning of the rate increase process was extremely tense. Customers were unwilling to even consider a rate increase without getting their questions answered. They wanted to know how money was being spent, what the money was needed for, and what LADWP was doing to cut costs (Fig. 11.3).

Environmental groups had certain issues that they wanted addressed including renewable energy goals, increasing energy efficiency, and getting off coal. Community and nonprofit groups wanted to make sure that seniors and low-income customers were not going to suffer. Business groups wanted to make sure that strategic investments would provide economic development in the City of Los Angeles.

In 2012, top management decided that the only thing to do was to get everything out in the open. Let the public know where the money was going, how far behind LADWP was with regard to infrastructure maintenance, and how much money was needed to meet basic operating needs. Along with this, the options for strategic investments including energy efficiency were presented showing their specific impacts on rates and the customers were asked to respond with their priorities.

> *"Worried about rising rates?*
> *Here's the truth: our electricity rates will rise no matter what kind of energy we use. The cost of coal is rising dramatically all by itself. But as we switch to clean energy, the DWP must invest wisely in energy efficiency programs. Regardless of rate increases, energy efficiency has tremendous potential to keep our electricity bills down. Furthermore, jobs in the clean-energy economy have grown at three times the rate of other sectors. Sierra Club Angeles Chapter, LA Beyond Coal Collecting Business Endorsements, June 1, 2011, Paula Block-levor*

**Fig. 11.3** Quote from Sierra Club writer Paula Block-levor

Five months and numerous presentations, workshops, and community collaboration sessions later, the tide was turned. At public meetings, some members of the business and community groups who had been heavily involved in the process positively commented on the transparency and involvement of the community. One participant from the North Valley Community Collaboration Session noted at one of the workshops, "I like this process and hope the public input is taken seriously. We have to create our own future and I think it would be preferable if it included making LA a mecca for sustainability and green."

LADWP listened and the rate increase that was adopted included strategic investments in energy efficiency, renewable energy, and early replacement of coal, in response to customer input. Energy efficiency would be a strong part of the plan, with provisions for low-income and special-needs customers and a focus on generating jobs and economic development in the community.

LADWP set energy efficiency goals in its 2012 Power Integrated Resource Plan (IRP), and later that year, the Board of Commissioners adopted guiding principles for energy efficiency. These included the following:

- Achieve an energy efficiency goal of 10 % by 2020, with a "reach" goal of 15 % by 2020.
- Leverage energy efficiency as part of the strategy for eliminating coal from LADWP's energy portfolio.
- Contribute to greenhouse gas (GHG) emission reduction through reduced energy use.
- Promote energy efficiency programs for all customer sectors, targeting "hard-to-reach" customers (i.e., low-income residents, small businesses).
- Leverage programs to support jobs for the local workforce.
- Work collaboratively with partner agencies to reach a broad and diverse customer base.

In August 2014, the Board of Commissioners formally approved increasing the energy efficiency goal to 15 % by 2020.

## 11.3.1 Efficiency Solutions Portfolio

LADWP began a concerted effort in June 2012 to update and consolidate its energy efficiency programs into a larger Efficiency Solutions Portfolio that utilizes an integrated customer-focused approach and partnering with other agencies and utilities to meet these new, aggressive goals. The Efficiency Solutions Portfolio includes joint programs that address both a customer's energy and water usage and individual programs that solely address energy use. LADWP also forged an agreement with Southern California Gas Company (SoCalGas) to provide combined electricity, water, and natural gas efficiency programs. Enhancing existing programs and developing new programs is an ongoing effort.

LADWP's overarching energy efficiency strategy utilizes a variety of delivery channels and partners to offer something to all of its residential, commercial, industrial, and institutional customers. Specific strategies vary somewhat relating to each of the programs, but generally include the following:

- Integrate energy solutions, as applicable, including water, electricity, and natural gas (via partnership with SoCalGas);
- Coordinate with statewide IOUs and POUs, as appropriate, to ensure an appropriate level of consistency to benefit those customers that have facilities in multiple utility service territories;
- Leverage existing delivery channels, as appropriate, to ensure a cost-effective portfolio that benefits LADWP customers;
- Leverage market partners (manufacturers, vendors, retailers, architects, energy professionals, community-based organizations, energy service companies, etc.) to market and deliver programs in a cost-effective manner.

The programs included in the portfolio are designed to address the diverse needs of LADWP's customers and to provide social, economic, and environmental benefits to the region. While each program provides multiple benefits, the table below provides a general grouping of the energy efficiency programs with respect to social, economic, and environmental sustainability principles (Table 11.1).

**Table 11.1** Three Pillars of Sustainability via LADWP Energy Efficiency Programs

| Social | Economic | Environmental |
|---|---|---|
| Home Energy Improvement | Small Business Direct Install | Emerging Technologies |
| Low Income Refrigerator Exchange | Commercial Lighting Efficiency Offering | Codes, Standards, and Ordinances |
| California Enhanced Home Program | Retrocommissioning Express | Refrigerator Turn-in and Recycling |
| Outreach and Community Partnerships | Custom Performance | City Plants |
| LA Unified School District Direct Install | Commercial Refrigeration | Savings By Design |
| Efficiency Item Giveaways | Chiller Efficiency | Electric Vehicle Recharging Rebates |
| | Energy Efficiency Technical Assistance | |
| | Consumer Rebates | |
| | Energy Upgrade California | |
| | LADWP Facilities Upgrades | |

### 11.3.1.1 Social

An important goal of the portfolio was to reach out to all segments of LADWP's customer base. Several programs are included in the portfolio. They achieve this objective by reaching out to low-income and hard-to-reach customers and assisting the public schools in saving energy and water. The Home Energy Improvement Program and the Low Income Refrigerator Exchange provide direct benefits to low-income customers at no cost to them. The California Enhanced Home Program targets high-density new construction, including single- and multi-family high-rise buildings, raising the bar for energy efficiency, which extends these benefits to future residents of these structures. The Los Angeles Unified School District Direct Install Program brings energy and water efficiency to public schools throughout Los Angeles, ultimately freeing up money from energy and water savings that can be used for other purposes.

An example of a unique program that promotes social benefits is the Community Partnership Program. This is an advocacy program that strives to improve customer awareness among "hard-to-reach" customers of energy and water efficiency programs through grants and assistance to nonprofit, community-based organizations. This program leverages nonprofits' established community networks, where familiarity and trust enable meaningful communication, in languages and environments where these customers are most comfortable. The energy and water savings accrue indirectly through adoption of best practices and increased participation in LADWP programs such as the Home Energy Improvement Program, Small Business Direct Install, and Refrigerator Exchange Program. The social benefits of this program are twofold. It assists low-income and hard-to-reach customers in reducing their energy and water costs, and it builds the capacity of community-based nonprofits.

The Community Partnership Program offers $20,000–150,000 grants to local nonprofit organizations that are selected to work in each Los Angeles City Council District or on a citywide basis. For an organization to participate in the program, it must have 501(c)3 status, be located in the City of Los Angeles, have an established track record of providing services to the community, demonstrate commitment to energy efficiency and water conservation through its existing programs, and have the capacity to track project progress, manage costs, and maintain financial records acceptable for a city financial audit.

Grant recipients are expected to complete their projects within twelve months. During the course of the year, a peer facilitator grantee provides support services to the other nonprofits, assisting them with the structure and evaluation of their programs, disseminating relevant information, and holding quarterly meetings where outreach methods are presented and where participants can exchange ideas and best practices.

LADWP has been able to fund a group of diverse projects ranging from introducing the topic of energy efficiency all the way to deep dive workshops. Projects vary and include youth-oriented projects, training of ambassadors, door-to-door canvassing, creative messaging, and curriculum development. Projects also focus on single- and multifamily residences and feature neighborhood forums and

**Fig. 11.4** SCOPE "Green the Block" event assisting residents in applying for LADWP energy efficiency programs

community events where electric and water efficiency devices are sometimes provided and low-water-use landscaping is discussed. Other projects focus on businesses, including small business outreach and targeted sector-based workshops (e.g., hospitals or bankers). Outreach is delivered in multiple languages through newly developed collateral materials, focusing on those languages that are most common in Los Angeles communities (Fig. 11.4).

One of the recipients in the program is SCOPE (Strategic Concepts in Organizing and Policy Education), a local nonprofit organization focused on building a better quality of life for low-income communities of color in South Los Angeles. This is what Elsa Barboza, the project manager for SCOPE, has to say about the program:

> "The LADWP has really wonderful programs, gems really, but nobody knows about them. People talk about behavior change on the part of customers to save energy. Behavior change is also needed on the utility side in getting information out to people. I applaud LADWP for testing out different ways of reaching people through this program. We have taken a dynamic and diverse approach to outreach; door-to-door with teams from the community who are familiar to the customer and speak the language; events like 'Green the Block' where we walk people through the process; and follow up support to help people when they do apply.
>
> This has benefited the community by helping people take advantage of programs that they would otherwise not have been able to either through not knowing about them or through reticence to tackle the process. Through SCOPE the community now has a voice at LADWP and SCOPE has the opportunity to do something positive for its communities."

### 11.3.1.2 Economic

Programs that address the economic pillar of sustainability encourage economic development, assist businesses, or promote local job growth. Several programs are included in the portfolio that specifically address commercial, industrial, and institutional customers. These range from a direct install program that targets small- and medium-sized businesses with simple energy efficiency needs to a custom performance and technical assistance program that targets larger, more sophisticated customers. These programs reduce operating costs for businesses and expand local economic growth in energy efficiency and related businesses. The programs that address residential customers, Consumer Rebates and Energy Upgrade California, promote economic activity in the residential construction and appliance markets. The LADWP facilities upgrade program controls operating costs for the utility, which ultimately pass through as savings to customers (Fig. 11.5).

One of the newer programs, the Energy Efficiency Technical Assistance Program (EETAP), promotes economic development and job growth by incentivizing investment in energy efficiency measures and technical services. EETAP goes a step beyond the assistance offered by standard programs. EETAP was designed to assist commercial, industrial, and institutional customers in closing the gap between project development and implementation for more complex building systems. By providing incentives for project development services including energy auditing and project management, LADWP aims to help its customers to strategically plan, follow through, and realize energy savings in the most cost-effective manner.

EETAP is targeted to LADWP customers that may be planning multiyear capital programs. EETAP provides funding to LADWP customers for professional energy efficiency services performed by participating engineering firms including the following:

- Comprehensive energy audit including analysis of lighting, HVAC, envelope, and process/plug loads;
- Identification of feasible energy efficiency measures (EEMs) and water conservation measures (WCMs) including life cycle cost analysis;

**Fig. 11.5** LADWP Headquarters, Downtown Los Angeles

- Financial analysis including identifying and calculating financing options, return on investment, and appropriate tax and utility credits/incentives for each measure;
- Preparation of a prioritized list of applicable measures based on feasibility, life cycle cost, and financial analysis;
- Preparation of a standardized energy audit report.

Past attempts to provide incentives for energy audits and assessments were unsuccessful in the actual implementation of measures. The current EETAP ties incentives to customer implementation of selected measures. EETAP incentive payments are processed after the implemented measures are verified and LADWP has received all required documentation.

EETAP includes a Trade Ally and Trade Professional Program to provide a selection of firms that meet certain minimum standards and have received training in the EETAP process and program guidelines, payment criteria, determination of incentive caps, and proper completion of the EETAP application package. The program incentivizes contractors and engineering firms to maximize implementation of cost-effective measures as a condition of participation in the Trade Ally and Trade Professional Program. The program also spurs economic development for engineering and professional firms by providing rebates for their services.

### 11.3.1.3 Environmental

Programs that address the environmental pillar of sustainability include programs that reduce air pollutants and greenhouse gas emissions such as the City Plants Program, the Refrigerator Turn-In and Recycling Program, and the Electric Vehicle Recharging Rebates. These programs also contribute to the advancement or widespread market penetration of energy efficiency such as the Emerging Technologies Program, the Savings By Design (SBD) Program, and the Codes, Standards, and Ordinances Program.

The SBD Program promotes the penetration of energy efficiency into new construction and promotes the advancement of energy-efficient building technology. This is a California statewide non-residential new construction program, in which LADWP partners with Southern California Gas Company (SoCalGas) to offer a uniform, multifaceted program designed to consistently serve the needs of the commercial building community. SBD encourages energy-efficient building design and construction practices, promoting the efficient use of energy by offering up-front design assistance, owner incentives, design team incentives, and energy design resources.

In addition to the traditional sliding scale incentives that are calibrated to energy savings exceeding standard energy performance code, SBD offers a kW reduction as well as financial support for design teams to undertake an integrated design process. Additionally, sustainability incentives are offered to building owners to achieve green building certification, perform building commissioning during design and construction, and establish and follow a building measurement and verification (M&V) plan after occupancy. These sustainability incentives are designed to encourage new buildings to be as well designed as possible, be built as well as they are designed, and be operated as well as they are built.

The primary objective of the SBD Program is to integrate energy efficiency into the design of buildings when more advanced measures can be implemented at a lower cost than for retrofit projects. Implementation of energy efficiency in the design process also enables taking a whole building approach in a manner that is much easier and cheaper than after the building has been constructed. The combination of education, design assistance, and incentives assists LADWP in further introducing and normalizing integrated design practices, which speeds the introduction of additional and incremental energy efficiency standards, further moving the non-residential new construction market to higher energy efficiency performance.

## 11.4 Internal Facilities Program

After LADWP developed a robust program for its customers, it was time to look at internal operations and make sure we were "walking the talk." The customer-based portfolio included a program for energy efficiency upgrades for LADWP facilities with an allocation of resources commensurate with other programs in the portfolio. What was needed was a comprehensive program that would enable implementing energy efficiency measures in a well-thought-out manner in order to achieve internal energy efficiency goals. Where to begin?

The primary focus of operations at LADWP is and has been to provide its customers with safe, reliable, and affordable electricity and water while maintaining environmental stewardship. Energy efficiency has always been embedded in the external services the utility provides. Bringing these practices to internal operations involved creating a structure for conservation within LADWP. There are four major thrusts of the program:

- Assess current state of facilities—Analyze, benchmark, and establish metrics;
- Engage internal stakeholders—Green Team, employees, executives, and managers;
- Set policy and procedure—Set targets, metering, and training and embed commitment;
- Make it happen!—Define projects, allocate internal resources, execute, and ingrain into policies/procedures.

The commitment must come from the top of organization to the shop floor for facilities to maximize energy efficiency.

## 11.4.1 Assess Current State of Facilities

LADWP's water and power facilities include 90 reporting locations with approximately 360 major buildings along with over 400 structures that serve system components such as receiving, switching, and distributing stations, and microwave and radio sites. The facilities energy management program focuses on the structures associated with the 90 reporting locations. Energy efficiency among system components is managed separately.

Building vintages range from the early 1920s to the present day. Most were built in periods when energy was inexpensive and not considered a major cost factor. In the late 1980s, energy efficiency began taking on more importance and energy efficiency upgrades started in LADWP facilities. In 2012, LADWP formed an internal Green Team made up of employees from various divisions. What began as a grass roots effort with a focus on improving the sustainability of the organization quickly became a vehicle for change. The group was enhancing LADWP's existing programs, developing new programs, and extending sustainability into general operations. In early 2013, the Green Team realized that a more concerted effort would be needed to expand sustainable practices, including energy efficiency, into all of its facilities and initiated the development of a sustainable facilities management program while engaging facility managers.

A pilot study was conducted to analyze energy use in LADWP facilities and to develop energy efficiency metrics. The purpose of the project was to develop metrics that could be used to:

- Assess the current state of energy efficiency for LADWP buildings compared to each other and compared to industry standards such as Energy Star, best management practices, and current codes.
- Provide a reliable source of information that could be used by management for prioritizing facilities for energy efficiency investments.
- Assess the actual performance of improvements with regard to energy and cost savings.
- Assess existing and reduced greenhouse gas emissions related to facilities operations.
- Provide a consistent resource for ongoing measurement and monitoring, leveraging existing reporting mechanisms.

The facilities chosen for the study encompassed a wide variety of operations that were common within the organization's facilities. Uses included office, fleet maintenance, equipment warehousing and distribution, customer service, equipment testing, cafeteria, data centers, electric vehicle charging, and emergency call centers. The energy efficiency analysis in the study included electricity use intensity (EUI), time-of-use, interval metering, potential energy and cost savings, and greenhouse gas emission reduction.

EUI metrics were developed by obtaining the EUI for buildings of categories similar to LADWP buildings utilizing the California Energy Commission

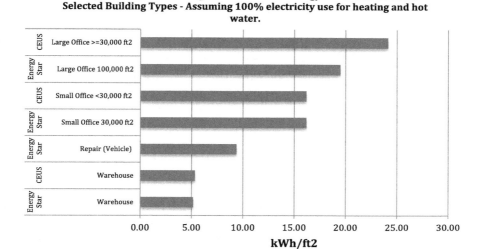

**Fig. 11.6** Energy Use Intensity (EUI) for Selected Building Types

(CEC), Commercial End-Use Survey (CEUS) Report (March 2006), and the US EPA Energy Star Target Finder. These metrics were to be used for screening purposes only. It was acknowledged that conducting a complete Energy Star Portfolio Manager analysis would provide a more accurate result, which could be different than the screening would indicate. However, the screening metrics provide a quick method of assessing a large portfolio of buildings using readily available data (Fig. 11.6).

While EUI and Energy Star Portfolio Manager provide useful information, time-of-use data enable taking the analysis one step further. LADWP has three billing periods, high peak (1:00–5:00 p.m. weekdays), low peak (10:00 a.m.–1:00 p.m. and 5:00–8:00 pm weekdays), and base (8:00 pm–10:00 a.m. weekdays and 24 h per day on weekends). Energy usage for each of these periods is reported separately on utility bills. Dividing the energy use for each period by the number of hours in the period yields the average energy demand for the period, a useful analytical factor (Fig. 11.7).

Graphing the average demand data shows a clear picture of energy usage throughout the year by time-of-use. It is clear from the graph below that energy usage during the nights and weekends is only slightly less than that during the daily peak periods, which represent weekday morning, afternoon, and evening hours. This suggests either a 24/7 operating schedule or that systems are not being turned back during non-work hours. As it turned out, part of this building was on a standard workweek and part on a 24/7 operating schedule.

The situation was that two of the HVAC systems served both types of areas, so they were running 24/7. The solution was to purchase separate split systems

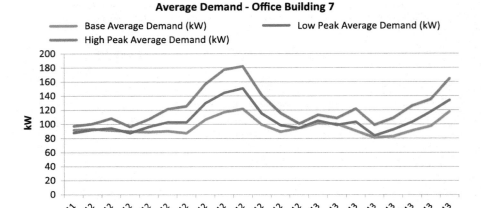

**Fig. 11.7** Average Electricity Demand for LADWP's Operations Building

for the smaller 24/7 areas and manage the package units on a standard workweek schedule. This change along with other recommended changes would result in a large decrease in energy demand during off-hours and a large decrease in energy use overall.

It was clear that a comprehensive analysis of energy and water usage for LADWP buildings could highlight inefficiencies and enable prioritizing solutions. Also, as discussed later in this chapter, the results showed that in some buildings, LADWP had achieved significant savings over the study period and that ongoing analysis would provide measurement and verification of savings.

### *11.4.2 Engage Internal Stakeholders*

It was clear that opportunities were available; the next step was to engage internal stakeholders to make it happen. There was a valuable range of experience and interest in the various divisions of LADWP, but roles and responsibilities were dispersed. LADWP's core function is to provide reliable energy and water services to its customers, and the organization is structured accordingly. So the challenge became how to integrate the human resource needs of an energy efficiency management program with the formal and functional structure of the organization.

Meetings were held with key people from various divisions to find out what roles they played, what resources they might be able to provide, and what functions they might be able to take on in this energy efficiency process. Meetings were held in engineering, architecture, IT services, finance, rates, energy efficiency, real estate, executive management, facilities management, maintenance, and others. Each meeting led to new connections with people in other divisions.

The idea initially met with a range of reactions from blatant skepticism to enthusiasm. The work of persuasion with different units began with reframing the approach, understanding what was of value to the manager, articulating benefits to LADWP and to individual asset groups, and finding ways to synchronize energy efficiency with existing employee roles and responsibilities.

Managers, facilities support personnel, and employees were engaged collectively and in separate meetings. Progress and queries were fielded to executive management to get their feedback and support. Up, down, and across, the goal was to get stakeholders to understand what we were working toward and to incorporate their ideas to address their concerns. In other words, getting to yes!

As the meetings progressed, a framework for the facilities sustainability program began to emerge. Approximately four months into the process, things reached the tipping point from persuasion to participation. Stakeholders began reaching out to the team with projects they were planning, facility managers in the pilot study were eager to proceed for their facilities, and other divisions began to see how this program could contribute to achieving their objectives.

### *11.4.3 Set Policy and Procedure*

The next step was to set policies and develop procedures for managing the implementation of energy efficiency at LADWP facilities. This involved setting energy efficiency goals for LADWP facilities and developing a strategic energy efficiency management plan that integrates energy efficiency projects with maintenance and capital improvement projects. Water efficiency would also be included as a result of a separate water study. The master facilities energy efficiency plan includes procedures for documenting existing facility equipment and conditions, identifying and ranking proposed energy efficiency projects, and designing projects to integrate multiple objectives such as energy, water, renewables, demand response, maintenance, and worker comfort, factors to consider in cost–benefit analyses, and potential sources of funding both within and outside of LADWP.

Additional policy changes that were recommended included the following:

- Give facility managers access to data for their facilities. Enable them to take a more active role in energy efficiency and water conservation for their facilities by providing them with information, tools, and training.
- Explore models to enable facilities managers to utilize part of the operating savings to invest in additional efficiency projects, eventually enabling them to develop energy and water management plans for their facilities.
- Include energy and water in facilities budgets and issue bills to the facilities so that they can see the need for and the results of their efforts.

## 11.4.4 Making It Happen

In order for this to work, the commitment and coordination of resources from multiple divisions across the organization would be needed. Another pilot program was conducted. A facilities sustainability management plan was prepared, projects were identified and ranked, and implementation of selected projects began. The key to the success of this pilot program was that it started with people who wanted to do it.

LADWP's commitment to the program became evident when the Green Team held a Sustainable Facilities Workshop for LADWP employees in late 2014. The workshop was opened by Nancy Sutley, LADWP's first ever Chief Sustainability and Economic Development Officer, and William Funderburk, the Vice President of the Board of Commissioners. The speakers included other members of LADWP's executive team, LADWP experts, and the facilities management team that participated in the pilot study. The workshop was attended by over 100 LADWP employees including facilities managers, engineers, architects, accountants, landscape maintenance personnel, and others and included a roundtable session that provided a great opportunity for employee input. Additional workshops are planned.

## 11.5 Results, Impacts, and Benefits

### 11.5.1 Job Creation in Los Angeles

#### 11.5.1.1 Results

The UCLA Luskin School of Public Policy conducted a study on the job creation benefits for LADWP's 18 largest energy efficiency programs in 2014. The energy efficiency programs ranged from 5.7 to 39 job-years per million dollars invested by LADWP, with a weighted average of 16 job-years per million dollars invested.[2] This compares favorably to other energy-related investments such as natural gas, smart grid, and solar generation as well as investment in residential construction, an oft-cited source of local employment growth. By the end of 2020, LADWP's energy efficiency programs have the potential to create nearly 17,000 job-years in Los Angeles (Table 11.2).

### 11.5.2 Energy and Cost Savings—LADWP

LADWP's headquarters, the John Ferraro Building (JFB), is a good example of the results of continuous energy efficiency investment and improvement. The JFB facilities team had been making energy efficiency improvements to this building

---

[2]DeShazo et al. [1].

**Table 11.2** UCLA Luskin School of Public Policy Study

| Investment Sector | Job-Years per Million Dollars Invested |
|---|---|
| LADWP Energy Efficiency Programs | 16 |
| Natural Gas | 5.2 |
| Smart Grid | 12.5 |
| Solar Generation | 13.7 |
| Construction | 10.7 |

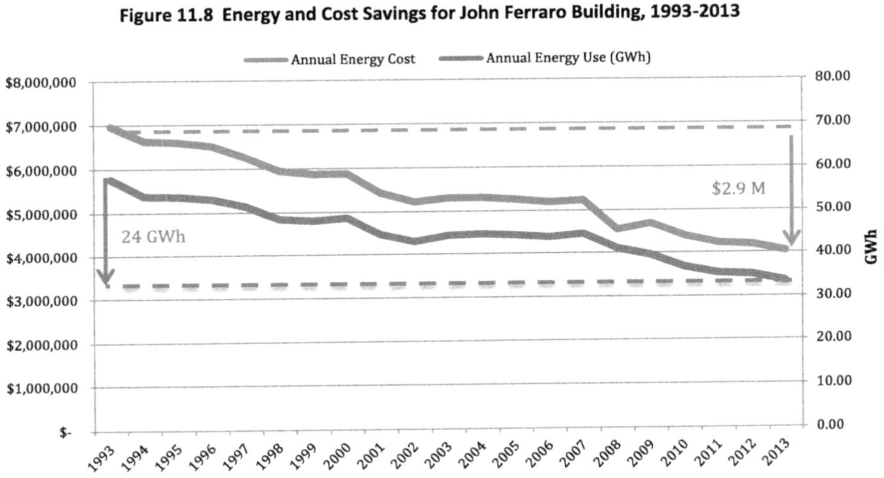

**Fig. 11.8** Energy and Cost Savings for John Ferraro Building, 1993–2013

since 1993. These efforts have reduced energy usage by 42 % for a cumulative energy savings of 24 GWh per year and a cumulative annual cost savings of $2.9 M per year. The benefits of these savings are reduced operating costs, which ultimately affect rates and a lowering of associated greenhouse gas emissions by 12,900 metric tons of $CO_2e$ per year. In June 2015, JFB celebrated its 50th anniversary and achieved Leadership in Energy and Environmental Design (LEED) Certification for Existing Buildings (Fig. 11.8).

## 11.6 Lessons Learned and Transferability

Energy efficiency plays a key role in sustainability for municipally owned utilities. LADWP's comprehensive portfolio of external energy efficiency programs enables customers to manage their energy usage and costs, includes programs for all customer sectors, creates local jobs, and reduces greenhouse gas emissions associated with power generation. LADWP's internal energy efficiency program reduces operating costs, engages employees in the process, reduces greenhouse gas emissions, and improves the performance of LADWP facilities.

## 11.6.1 Importance of Engaging Stakeholders

Today's environment of heightened community interest requires municipally owned utilities to obtain stakeholder input on major issues. Stakeholders want their voices to be heard and concerns addressed. When facing rate increases, engaging stakeholders in the process by providing them with a clear explanation of alternatives and soliciting their feedback helps shape solutions and gain acceptance. Through its stakeholder engagement, LADWP has developed a coalition of labor, business, and environmental groups in support of energy efficiency. Without that coalition, LADWP would not have the third largest energy efficiency portfolio in California and would not have obtained the rate increase needed to support it.

## 11.6.2 Internal Energy Efficiency Management

Formalizing energy efficiency management is a challenge for any organization, including municipally owned utilities. Resistance to change is a large barrier to any new program. Start with the people in the organization who want to do it; others will follow when they see positive results. Consider the diverse needs and concerns of involved departments and identify their budget and resource gaps. Address multiple needs by integrating energy efficiency with demand response and operations and maintenance objectives. When it comes to energy efficiency, the traditional focus of municipally owned utilities has been on customer programs. Recognize that it will take a transformation to turn the focus inward.

## 11.7 The Next Decade

Energy efficiency is still in its early stages. Several trends are occurring that will accelerate growth in energy efficiency and change its nature. Rising electricity rates, increasing interest in sustainability, state and federal policies, and emerging technologies are all combining to accelerate the implementation of energy efficiency in residences, buildings, and businesses. The influx of electric vehicles and the increase in electronic devices are transforming the way we look at energy efficiency.

On the customer side, as energy costs rise and energy demand increases, LADWP is taking a more aggressive approach to energy efficiency for its customers. Energy efficiency goals have been increased to a 15 % reduction in energy usage by 2020. New and innovative energy efficiency programs are being developed that target customer needs yet maintain overall cost-effectiveness of the portfolio and plans are underway to address emerging market trends.

On April 8, 2015, the City of Los Angeles released its Sustainable City pLAn. Energy efficiency is a key component of the plan, which includes strategies and key initiatives to achieve near-term and long-term energy efficiency objectives on a citywide basis. LADWP has been working closely with the city in the development of the plan, and LADWP programs and policies will continue to influence and be influenced by the implementation of the Sustainable City pLAn.

Rapid advancements in analytics technology are transforming the energy efficiency market. Utility smart meters and new cloud-based energy management software are enabling an intelligent, broad-based approach to energy efficiency.[3] Utilities can actively utilize energy performance monitoring to conduct remote audits, track savings over time, and ensure continuous feedback from customers, enabling sustained, persistent savings that the utility can rely on to meet its goals. LADWP is set to deploy the nation's largest analytics-enabled efficiency program in which 70 different energy efficiency parameters will be assessed through advanced analytics to rapidly target 2000 commercial buildings for remote audits and recommendations for energy savings.

Electrification of the vehicle fleet is adding a whole new dimension to energy use and the way we look at energy efficiency. Electric vehicles are leading to the intertwining of land use and transportation into what has become known as the "land use—transportation nexus." As an ever-larger fraction of the vehicle fleet becomes electric, LADWP will have a direct interest in measures and programs that reduce vehicle miles traveled (VMT), such as carpooling, public transportation and supporting compact, mixed-use development patterns that reduce the need for personal vehicle travel.

Internally, LADWP will be leading the way by "walking the talk." LADWP is taking a more aggressive approach to the energy efficiency of its own facilities and operations and is taking a more active role in supporting emerging technologies.

In 2015, LADWP was the first electric and water utility to join the US Department of Energy's (DOE) Better Plants Challenge. Under this program, LADWP's Water System has committed to reducing the energy intensity of its facilities by at least 25 % over the next 10 years. Achieving these goals will take a culture transformation within the organization to make energy efficiency an integral component of operations and will involve employees from all sectors of the organization.

Located in the Arts District area of downtown Los Angeles, the La Kretz Innovation Campus opened in late 2015. This facility serves as a clean industry hub for entrepreneurs, engineers, scientists, and policymakers to interact and promote the development of clean technologies and Los Angeles' green economy. LADWP's Energy Efficiency and Emerging Technology Center is also housed at the facility where new technology and equipment can be developed and tested, accelerating their introduction and penetration in the commercial market. In addition, the project is located in a designated Enterprise Zone where it will help revitalize a low-income community through economic development and job creation (Fig. 11.9).

---

[3]Gruenich and Jacot [2].

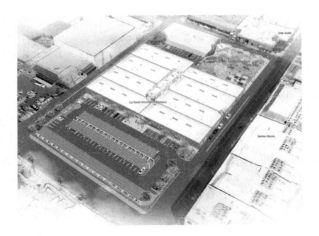

**Fig. 11.9** La Kretz Innovation Campus

## 11.8 Conclusions

As a sign of the increasing importance of sustainability to LADWP, Nancy Sutley was appointed as the first Chief Sustainability and Economic Development Officer in 2014, bringing with her extensive experience from her leadership role as the Chair of the White House Council on Environmental Quality. She has this to say about sustainability and municipally owned utilities, "The landscape is changing. As sustainability gains traction in the private sector, municipally owned utilities are constantly monitoring policies and practices to keep pace with the principles of sustainability. These changes require a change in the way of thinking; looking internally as well as externally and involving external as well as internal stakeholders. While the process may be more involved, the results are more resilient and provide opportunities for innovative solutions. LADWP is committed to being a national leader in energy efficiency and water conservation, both for its customers and its own operations."

## References

1. DeShazo JR, Turek A, Samulon M (2014) Efficiently energizing job creation in Los Angeles. UCLA Luskin Center for Innovation, Los Angeles, July 2014
2. Gruenich D, Jacot D (2014) Scale, speed, and persistence in an analytics age of efficiency: how deep data meets big savings to deliver comprehensive efficiency. Electr J 27(3):77–86. Available at: http://dx.doi.org/10.1016/j.tej.2014.03

# Chapter 12
# Southern California Edison: Renewable Energy

**Dawn Wilson, Ron Gales and Pat Adams**

**Abstract** SCE has been researching and investing in renewable energy sources such as solar, wind, geothermal, and hydropower since the 1970s. This commitment to renewables stemmed from federal and state regulations aimed at reducing air pollution and—more recently—greenhouse gas emissions as well as the need to meet the growing capacity requirements of California's population, which has grown steadily for decades. SCE sees competitive pressures, new technologies, and aggressive environmental policies where the energy sector is expected to contribute significantly to the reduction of greenhouse gases as prominent drivers of industry change in the coming years. The company will continue to adapt its business strategy to navigate through these changes, leveraging its experience and skill in pioneering new technologies to re-envision a flexible, resilient grid that enables customer choice and clean, sustainable energy.

## 12.1 Introduction

"It is the policy of the Southern California Edison Company to devote our corporate resources to the accelerated development … of future electrical power sources which are renewable rather than finite," declared SCE's Chairman and CEO William R. Gould at a press conference in downtown Los Angeles. "These include

---

D. Wilson
Environmental Policy and Affairs, Southern California Edison Company, Rosemead, USA

R. Gales · P. Adams (✉)
Southern California Edison Company, Rosemead, USA
e-mail: patricia.adams@sce.com

wind, geothermal, solar, fuel cells, and continued emphasis on co-generation, conservation and load management." The date was October 17, 1980.

Gould's actions were influenced by the passage of PURPA (Public Utility Regulatory Policies Act of 1978). PURPA was passed as part of the National Energy Act—constructed in light of the first energy crisis of 1973. PURPA was designed to promote electricity and natural gas conservation, energy efficiency, and development of small hydroelectric facilities (in addition to its provisions for crude oil transportation).

The term "renewable" has emerged over the last decade as a byword in the global conversation about energy, yet as this policy announcement illustrates, SCE's involvement with renewable energy resources such as solar and wind reaches back more than four decades—even longer when considering geothermal energy and hydroelectric power.

The push for renewable generation in energy markets worldwide is rooted in the desire to reduce greenhouse gas emissions and to realize the vision of a sustainable energy future. Clearly, the growing importance of renewable energy is one of the drivers of transformative change in the electric power industry. There are several long-term trends that we expect will continue to significantly influence the industry, most important of which are as follows:

- Public policy prioritizing environmental sustainability;
- Technology and financing innovations facilitating conservation and self-generation;
- Regulation supporting new forms of competition;
- Flattening domestic demand for electricity.

SCE ("Edison") believes that addressing global climate change is an important issue. We are committed to meeting our obligations under California's ambitious energy legislation of recent years (*see* "California GHG Policy," below). In addition, we welcome the opportunity to work with the US Environmental Protection Agency, the administration and Congress to develop federal policies that would align with California's programs and reduce greenhouse gas (GHG) emissions without adversely affecting our ability to provide reliable and affordable electricity to our customers.

Serving our customers is a top priority. Ted Craver, the present-day chairman and CEO of SCE parent company Edison International, said in a recent speech, "What our customers and our communities want … [is] an electric system that is modern, that will be environmentally responsible and allow California to stand in the front in terms of carbon reduction and greenhouse gas reduction. So our role is to help our customers and policy makers understand the tradeoffs that are involved but once they understand those tradeoffs, our job is to go out and build the system that will meet those objectives."[1]

SCE's core mission remains the same since the company was founded: to safely provide reliable and affordable electric service to our customers. Today, our company is augmenting the power grid in our service territory to better

---

[1] Ted Craver, "Keynote" (Institute for Corporate Counsel, Los Angeles, CA, December 3, 2014).

incorporate energy from solar, wind, geothermal, and other sources. Energy diversity protects customers from price fluctuations, and it reduces our dependence on any one source. An "all-of-the-above" approach to energy sources will give California the reliability and affordability our customers depend on.

## 12.2 Policy Decision

In the 1980 announcement, Gould explained that Edison projected a load growth of 11,000 MW for the decade beginning in 1980. With the anticipated success of conservation and load management, the estimate had been reduced to 6000 MW of additional generation capacity necessary to serve projected customer needs for the coming decade.

SCE committed to more than doubling its targeted generation capacity from renewable and alternate energy sources, from 800 to 1900 MW by 1990. This was a greater commitment than that of any other utility in the nation at that time. A *Washington Post* editorial on October 29, 1980, hailed Edison for "a courageous and welcome change of course…it may well have a greater impact on other utilities' thinking and planning than all the government and academic studies combined." The *Wall Street Journal*'s Laurel Leff wrote that SCE "may have signaled a significant change in the direction for public utilities and U.S. energy production."[2]

At the time of Gould's renewable announcement, the company's reputation for innovation was already long-established. In 1899, SCE precursor Edison Electric began transmitting power to Los Angeles over the world's longest power line to date—83 miles. That mark was eclipsed in 1907 when Edison Electric's Kern River-Los Angeles Transmission Line began operation as the world's longest (118 miles) and highest voltage (75 kV) power line.

In addition, SCE had long championed clean energy technologies. In 1947, the year that Los Angeles formed the nation's first Air Pollution Control District, SCE established the first pollution control program by a utility in the USA. During the 1970s, the company devoted about 20 % of its capital budget to the installation of pollution control equipment on its coal- and natural gas-fired plants.

In this context, SCE's 1980 policy decision—decades before "renewable energy" joined the national lexicon—represents a natural progression for a company with a history of innovation.

Most importantly, committing to "accelerated development" of renewable energy was a business decision with long-term implications, rather than solely a reaction to pressures from regulators or environmental activists. In his remarks, Gould cited significant progress with "forms of power generation which a few years ago were speculative or unproven" as a key factor in the policy decision. The announcement marked a point in time when these alternative energy sources,

---

[2]*Edison News*, November 21, 1980.

combined with energy conservation, were becoming more technically feasible and more economically viable when compared to the climbing capital costs and extended licensing timelines of new coal-burning or nuclear plants.

That said, the oil shocks and energy crises of the early and mid-1970s were still fresh in the nation's consciousness, and advocacy groups such as the Environmental Defense Fund and the National Resources Defense Council—descendants of the grassroots activism of the 1960s–1970s environmental movement—were becoming more mainstream and gaining influence. Gould tacitly acknowledged these factors when claiming that SCE's new policy "should improve the environment, should reduce dependence on expensive foreign oil, and should generally improve the air quality in the South Coast Air Basin."

## 12.3 Renewables: Research and Development

SCE could make such a policy decision only with the confidence it earned through its research and development successes in the years leading up to the announcement and the road map that resulted from those successes, for example:

*Solar:* In 1976, the US Energy Research and Development Administration (ERDA) selected SCE's Cool Water plant in the Mojave Desert as the site for the nation's first large-scale solar electric generation site, later named Solar One. The 10-MW solar power station, constructed and operated by SCE in partnership with the US Department of Energy (DOE), the California Energy Commission (CEC), and the Los Angeles Department of Water and Power (LADWP), began operation in 1982.

Solar One was the largest plant of its type in the world. It was the first commercial plant in the USA to use solar thermal technology, which generates power by focusing sunlight on a central receiver filled with an oil medium; the heated oil circulated through a heat exchanger, flashing water into steam and driving a turbine.

It included a thermal storage subsystem capable of providing steam to generate up to 7 MW of power during temporary periods of cloud cover and in early evening hours. Solar One supplied energy until 1986, when it was redesigned as Solar Two, which produced power from 1996 to 1999. Using molten salt rather than oil as a medium, Solar Two was more efficient than its predecessor and could produce electricity long after sunset.

*Wind:* SCE also took part in ERDA's wind energy program and was an early leader in wind power. In 1978, SCE opened a wind research center next to its Devers Substation near Palm Springs. In the early 1980s, many manufacturers were invited to field test early designs of wind-powered generators; these included everything from a Dutch windmill-style, wooden propeller machine built by Bendix to a 500-kW Darrieus vertical axis model, to an early version of the horizontal axis model that today has become a familiar sight.

Around the same time, SCE took steps to help third-party wind developers coordinate their investments in two areas with promising wind resources located in the Tehachapi Mountains southeast of Bakersfield and the San Gorgonio Pass west of Palm Springs. These areas contained more than 7200 wind machines by the end of 1985.

***Geothermal***: SCE's longtime interest in geothermal energy was rooted in simple geography; some of the world's best geothermal resources are located in SCE's service territory. As early as 1971, SCE had reinvigorated its investigation of geothermal power, looking at the Imperial Valley and the Mammoth Lakes area in the eastern Sierra Nevada Mountains and producing encouraging developments in both places in the mid- and late 1970s.

In 1978, SCE partnered with the Union Oil Company to build a geothermal plant at Brawley near the Salton Sea and a second plant shortly afterward at nearby Niland. SCE later ceased involvement with both projects, as operations became uneconomical. Third parties working at other projects in the Imperial Valley and Mammoth Lakes geothermal areas achieved commercial success, and these companies now sell power to SCE as part of the company's renewable energy portfolio.

***Hydropower***: SCE has long had a special connection to hydroelectric power, as the Big Creek Hydroelectric System in the Sierra Nevada Mountains—which another SCE precursor, Pacific Light and Power, began constructing in 1910—continues to represent a defining moment in company history. When energized in 1913, the 241-mile 150-kV line from Big Creek to Los Angeles set records for voltage and distance. The complex provides power to Southern California to this day, supplying 1000 MW of cost-effective, renewable, and environmentally sustainable power, enough to power about 650,000 homes.

By year-end 1980, hydropower remained the company's primary source of renewable or alternative power and represented 6 % of SCE's total generation capacity. Today, because the company divested its holdings over time in coal-fired and nuclear-powered generation facilities, hydropower accounts for about 20 % of SCE's utility-owned generation capacity. However, California policy does not include power from large (greater than 30 MW) hydro plants like Big Creek toward the state's Renewable Portfolio Standard (RPS) goals, yet *does* include small (less than 30 MW) hydro plants, as well as *purchased* renewable energy. Thus, geothermal, wind, solar, and biomass—both utility-owned and purchased—have overtaken hydropower as SCE's leading sources of renewable or alternative power.

## 12.4 California GHG Policy

California takes pride in being at the forefront of renewable energy and environmental policy. Part of that is due to the state's history of combatting air pollution, and part is due to the influence of Silicon Valley and the state's enthusiasm for attacking problems with technology. The early 2000s saw a flurry of statewide legislation and programs designed to further environmental sustainability. This spate of legislation marks the point in time when public policy became the key driver of SCE's renewable energy strategy.

In 2002, the California legislature enacted a bill establishing the California RPS, requiring electricity retailers in the state to purchase 20 % of their power from renewable sources by 2017. Subsequent legislation four years later accelerated the deadline to 2010.

California increased its renewable requirement late in the decade with a proposal requiring 33 % of delivered power to come from eligible renewable energy resources by 2020; this became state law in 2011 and is one of the most ambitious renewable energy standards in the country.

In 2004, two years after the California legislature had passed a law requiring electricity retailers in the state to purchase 20 % of their power from renewable sources by 2017, Governor Arnold Schwarzenegger announced the Million Solar Roofs Program. The goal was to install 3000 MW of rooftop solar photovoltaic (PV) statewide by 2016. Projections were that, compared to burning fossil fuels to produce the same amount of power, this would cut greenhouse gas emissions by three million tons—the equivalent of taking a million cars off the road. The legislature authorized the program in 2006.

The capstone of this legislative activity was the California Global Warming Solutions Act of 2006. This state law aims to reduce California's GHG emissions to 1990 levels by 2020. The California Air Resources Board's (CARB) Greenhouse Gas Cap-and-Trade Program is one measure to help achieve this goal. It places a statewide upper limit, or cap, on greenhouse gas emissions. Emissions are to be reduced by 2–3 % each year between 2013 and 2020 in order to reach the 1990 emissions level target. CARB allocated allowances to the electricity sector based on historical emissions and early action reductions. SCE's allocation is projected to decline from 33 million metric tons of carbon dioxide equivalent (MT $CO_2$e) in 2013 to 25 million MT $CO_2$e in 2020. These allowances will offset some of the GHG costs associated with the electricity SCE generates and purchases to serve our customers.

It is fair to speculate the degree to which California's RPS Program and Global Warming legislation influenced the US Environmental Protection Agency's Clean Power Program, announced in June 2014. This program aims to cut carbon pollution from the power sector by 30 % from 2005 levels by assigning a target carbon emissions rate to each state to achieve by 2030, based on the state's level of carbon emissions from fossil fuel-fired power plants divided by its total electricity generation.

In addition, the California Public Utilities Commission has directed the state's three investor-owned utilities (Southern California Edison, Pacific Gas and Electric, and Sempra) to procure 1300 MW of energy storage by the end of the decade. Energy storage is crucial to successful integration of renewable energy; in theory, it will provide a method for injecting stored energy into the grid when the sun is not shining or the wind is not blowing. The role of utilities in California's fight against climate change is more important than ever before. In late 2015, the state legislature passed SB350, "the Golden State Standard," which dramatically altered the energy landscape in California. The bill, which codified into law a series of previous gubernatorial executive orders, requires an increase in renewable procurement by the state's utilities to 50 % in 2030, and a push to double existing energy efficiency goals. This legislation also created an Integrated Resource Planning process, in which environmental goals will be key inputs in utility procurement decisions.

SB350 states that a principal aim of "utilities resource planning and investment [will be] to improve the environment and to encourage the diversity of energy sources through improvements in energy efficiency, development of renewable energy resources (…) and widespread transportation electrification." With this clear legislative directive, California utilities and state regulators are diligently investigating ways that utility investment can continue to reduce harmful emissions in the both the energy and transportation sectors.

## 12.5 Clean Energy Programs

SCE has demonstrated its support for California's GHG emission reduction efforts by significantly investing in clean energy programs and practices.

In 2014, about 23.5 % of the energy delivered to SCE's customers was from renewable sources. Solar energy comprised 15 % of the renewable energy delivered. (Other renewable generation sources include wind, geothermal, biomass, and small hydro.) SCE is on target to achieve the state's 33 % RPS by 2020; to do so, it will increase its RPS eligible renewable purchases by 7.4 billion kWh, or 42 % from 2014 levels.

SCE has helped thousands of homeowners, businesses, and multi-family residences generate their own power using a range of solar technologies. According to the Solar Electric Power Association, SCE ranked second among utilities nationwide in 2014 for the number of solar customer interconnections, bringing 34,588 customer systems into the grid. That is a new solar customer interconnection every 15 minutes.

Governor Schwarzenegger's Million Solar Roofs Program exists today as the California Solar Initiative (CSI), a ratepayer-funded program that pays an incentive to customers who install photovoltaic or solar thermal systems on their homes or businesses. Since the CSI's inception in 2007 through the end of 2014, SCE has paid out more than $714 million in rebates to more than 62,000 customers. As of year-end 2014, there were 320.09 MW of installed residential solar (with 6.9 MW pending) and 341.1 MW of installed non-residential solar (with 210.3 MW pending).

Meeting the state's renewable energy goals means SCE must build new high-voltage transmission lines or upgrade existing lines. SCE is building the 173-mile Tehachapi Renewable Transmission Project Segments 4-11, part of the nation's largest transmission project devoted primarily to renewable energy (Segments 1-3 were completed in May 2010). It will deliver up to 4500 MW of power to California's grid—enough to power 3 million homes. In 2013, SCE completed construction on the 153-mile Devers-Colorado River (DCR) and 35-mile Eldorado-Ivanpah (EITP) Transmission Projects. The EITP project can now deliver up to 1400 MW of power from renewable and traditional generating sources, and DCR can accommodate up to 2300 MW, once the West of Devers upgrade is completed.

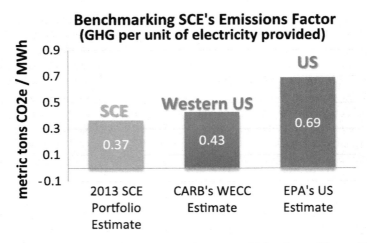

**Fig. 12.1** Benchmarking SCE's Emissions Factor. *Note* WECC refers to Western Electricity Coordinating Council

As of year-end 2013, SCE's GHG intensity per unit of electricity served remained well below the estimated national and western US averages (Fig. 12.1).

SCE continues to invest in researching and evaluating the clean, sustainable energy technologies of tomorrow. Today, at its Advanced Technology Centers in Pomona and Westminster, CA, the company researches and develops plug-in electric vehicle technology, battery storage, and smart grid applications that will allow increasing amounts of distributed energy resources (DER) to be safely and reliably connected to the grid.

## 12.6 Challenges to Meeting Renewables Targets

Meeting the company's renewable targets has not come without its challenges. Consistent with transition out of a significant portion of utility-owned generation assets, SCE has generally accomplished its renewable objectives through contracting with independent power producers (IPPs)—putting the utility in the role of market facilitator dependent on the actions of outside companies to deliver generation of the right size, at the right location, at the right time.

As stated above, California's RPS required electricity retailers in the state to deliver 20 % of their power from renewable sources, establishing an aggressive target for the state's utilities. This was subsequently increased to 33 % by 2020. SCE's preferred approach to further development of the renewable generation market is through an open competitive bidding process across broad markets. The aggressive targets led developers to propose many large projects which proved difficult to deliver in the time frames needed to meet the state's goals. According to Mike Marelli (currently Director, Strategic Planning, EIX Fiber Solutions, and formerly Director, Origination and Analytics), during the implementation of RPS the company had to learn how to combat a tendency to fill up the project queue with very large renewable projects that ultimately met the state's goals—but required the SCE team to simultaneously manage a number of very large, very complex projects to completion, all within aggressive time frames.[3] In the current Preferred Resources Pilot (*see* Preferred Resources Pilot section below), SCE modified its bidding process to allow smaller bids by renewable providers.

Renewable projects are complex in nature, involving a number of parties, many of which may have a concern with some aspect of the project. Resolving concerns raised by stakeholders can be time-consuming and often results in unanticipated delays and/or cancellations of a project. Over time, the SCE developers and other stakeholders have become better at defining and understanding criteria sooner in the process, which should increase the success rate of proposed projects. In response to the question, "What are some of the key lessons SCE has learned regarding ensuring project success?" Matt Langer, Principal Advisor, EIX responded, "Lessons learned include doing due diligence to ensure that each proposed project already has a completed interconnection study and the developer has secured site control and permits."[4]

Transmission availability and the time to study and perform any upgrades to support the delivery of renewable generation cannot be overlooked in siting projects; this has proven to be one of the biggest hurdles in meeting RPS requirements. Local communities impacted by larger renewable transmission projects are putting increased pressure on utilities for underground transmission lines that have typically been overhead. For example, in response to pressure from the City of Chino Hills, the California Public Utilities Commission (CPUC) issued a decision in July 2013, requiring SCE to underground a 3.5-mile segment of a 500-kV transmission line in the City of Chino Hills, part of Edison's Tehachapi Renewable Transmission Project (TRTP).[5] SCE evaluates, as part of the competitive selection process, the total expected costs to deliver the energy including the cost of transmission investments. However, if these cost expectations change, then the project selection and timing may still not be valid. This has been a key challenge to contracting—sharing that risk between SCE and the developers.

Key to the successful delivery of renewable projects is recognizing there are going to be stakeholder impacts regardless of the project. In each project, SCE has had to balance the environmental benefits associated with the project and the potential environmental impacts of project implementation. You cannot implement a renewable project without having some impact on the environment—e.g., habitat, species, and historical landmarks. There are always trade-offs that have to be taken into account. If you implement a solar project, you will impact ground species. If you implement a wind project, you will impact birds. There are a myriad of other potential impacts, for example to Indian burial grounds. These issues are even greater as we work to meet the transmission challenges noted above, where sizable generation needs to be sited far away from electricity use. Over the years,

---

[3]Personal Communication, April 7, 2015.
[4]Personal Communication, April 20, 2015.
[5]D.13-07-018, issued by CPUC on July 11, 2013.

SCE has learned to work effectively with stakeholders to discuss and develop proactive approaches to address these issues as early in the process as feasible.

## 12.7 Business Strategy Impacts

Clearly, emissions reduction and environmental responsibility is at or near the top of California's legislative and regulatory agenda. In this climate, SCE's challenge is to provide electricity to our customers in an environmentally responsible way that maintains safety, reliability, and affordability so as not to impede economic growth.

Increasing the use of renewables is a central component to any long-term solution. Yet utilities like SCE must realistically address the cost issue; at this time, the levelized cost per kilowatt for a utility-scale wind or solar farm can be two or three times greater than that of a natural gas-fired plant. The relatively quick decrease in recent years of solar energy production costs is encouraging for those who seek greater adoption of renewable energy, but production costs must continue decreasing significantly to achieve at least near-parity with conventional sources in cost per kilowatt.

From a reliability standpoint, the ability to manage intermittent resources like many renewables has significant implications, especially for commercial and industrial customers whose operations and processes are increasingly computerized. Even a momentary outage can jeopardize their computerized controls and require costly restoration. At a time when these customers need even greater reliability, utilities can scarcely afford service setbacks.

In analyzing this landscape over the years, executive leadership at SCE, and at its parent company Edison International, has evolved the company's business strategy to best position for growth. The company has scaled down its generation portfolio over time and put greater focus on its "wires" business, especially on growing opportunities in the distribution sector. Consistent with the company's heritage, SCE and Edison International are looking to technological innovation as key to balancing the competing demands for affordability, reliability, and environmental responsibility.

At the time of Gould's 1980 announcement, SCE fully owned all its generation assets. Since then, like many other utilities, SCE divested itself of many of those assets as the IPP market has developed capabilities to meet those needs. Today, SCE procures about 85 % of its power through contracts primarily from natural gas-fired plants and renewable energy producers. (Note: In December 2013, SCE completed the sale of its 48 % interest in Units 4 and 5 of the Four Corners Power Plant to Arizona Public Service Company. With the sale, SCE no longer owns any coal-fired generation.)

The company's growing focus on its "wires" business can be seen in its rising investment in infrastructure, primarily in the distribution or "small wires" sector. SCE's overall capital expenditures rose to the $3.8–$3.9 billion range in 2010–2012, up from $2.4 billion in 2008. Looking forward, SCE forecasts between $3.6 and $4.6 billion per year in capital expenditures from 2014 through 2017, pending regulatory approval. The majority of this spend is dedicated to replacing and updating aging infrastructure, much of which dates back to the post-WWII era, with what SCE calls the "twenty-first century power network."

Such a network integrates digital age technology into the electric grid to read, monitor, and process data and dynamically control devices in that system. The platform SCE envisions will help manage the dispatch of intermittent resources and mitigate the inefficiencies of how electricity is routed through a distribution system designed decades ago.

SCE's guiding principle for this vision is to enable two-way power flows between utility infrastructure and a range of generation sources, such as rooftop solar and energy storage, located as close as possible to the load being served. Collapsing the distance between generation and load served (*see* "Distributed Energy Resources" below) is a major step in realizing routing efficiencies and reducing emissions.

## 12.8 Distributed Energy Resources

While the electric power industry is clearly undergoing transformative change, the bulk power system is still the backbone of our industry. Traditional flow of power from centralized generation to load is likely to continue to play a significant role in serving electric system demand for the foreseeable future.

A strong bulk power system helps ensure a reliable supply of electricity and facilitates robust wholesale electricity markets. It also helps meet public policy goals by providing access to utility-scale renewables such as wind farms and solar-generating stations.

SCE views the bulk power system as not just central stations and transmission and distribution, but also resources including storage, demand response, and energy efficiency. All of these can be used to balance the load demand with intermittent generation.

That said, anyone flying into the airports of Southern California can catch a bird's-eye glimpse of the future of the electric power system. This vast region is dotted with the reflections from shiny solar panels on rooftops of homes, schools, and businesses. PV solar systems also can be found on some parking lots and warehouses.

PV solar is the most visible segment of a major, ongoing transformation of our electric system, known as distributed generation, or more broadly, distributed energy resources (DERs). These resources include power generators, storage, and demand response. The generation is typically smaller than 10 MW, located at or near customer sites—PV solar as well as natural gas-fired micro-turbines, combined heat and power systems, small wind turbines, and fuel cells. Storage is a growing part of DER and is expected to help SCE utilize variable renewable generation resources better. Demand response is also another DER "knob" we can turn to help SCE better balance generation resources and customer loads.

The challenges accompanying the growth of DERs are among the most complex questions that must be addressed if renewable energy is to become a larger portion of our nation's energy mix. For example, the distribution grid that SCE operates was designed for one-way flow of electricity from power plant to customer. However, DERs cause two-way flows when, for example, a customer's solar generator feeds power back into the system. That can cause fluctuations in voltage and frequency and make safe system operations more difficult, creating reliability problems if the distribution grid has not been modified to handle such flows. The variable nature of most renewable resources, especially rooftop solar, creates a challenge for our grid operators who must continuously and instantaneously manage supply and demand. SCE's plan to modernize the grid is designed to increase the capabilities for system planners and grid operators so that the challenges of variability can be mitigated and DERs can be fully integrated into the SCE system.

Affordability and equity issues are also involved. All residential customers, even those with rooftop solar panels, use the power network—for example, they all rely on the grid to supply electricity at night—so all these customers should help pay for their use of it. SCE believes that all residential customers should contribute to maintaining and upgrading the power network we all rely on with a flat fee that is separate from energy use and makes the cost of using the grid transparent. This flat fee will reduce the energy use charge and does not represent new revenue for utilities. One of the key principles of the twenty-first century power network is that all customers can have access to clean energy and efficient technologies.

## 12.9 Grid Modernization

Taking the lessons learned from SCE demonstration projects including the ARRA/DOE sponsored Irvine Smart Grid Demonstration[6] and laboratory testing of various DER devices, SCE is embarking on a grid modernization effort. This "modernization" of the grid will prepare the substations and distribution system for interconnection of renewable generation, storage, and new loads such as electric vehicle charging. The initial demonstration of these system upgrades will be implemented as part of the SCE Integrated Grid Project which will install equipment and control systems to allow coordinated operation of the distribution system with significant amounts of DER. These upgrades include devices to enable faults to be identified more quickly, sensors to provide better information to system operators, and more automated control of the distribution system operations. Underlying all of these changes will be an upgraded communications and cyber security system to enable safe and reliable operations. These changes to the SCE system will start now and take place over the next 5–10 years.

---

[6]A project summary and technical reports can be found at the following Web site: https://www.smartgrid.gov/project/southern_california_edison_company_irvine_smart_grid_demonstration.html.

## 12.10 Preferred Resources Pilot

Amid the promise of challenges of integrating renewable energy into the power grid, SCE is currently conducting a real-world test in a densely populated area of its service territory to explore the use of renewable energy and other low- or no-emission resources to offset local load growth. Clean, renewable and sustainable energy can come from several sources—in terms of generation (clean distributed generation), conservation (energy efficiency and demand response), and energy storage. These sources are collectively called preferred resources and may be a suitable option to meet local electrical demands. However, the use of these sources of power also means that the electricity grid must be upgraded to integrate more electric power sources than ever before.

SCE launched its Preferred Resources Pilot (PRP) in 2013 in South Orange County, where the company anticipates more than 250 MW of load growth over the next 10 years. Through existing and new procurement and customer programs, a mix of resources will be installed in the local zone and SCE will measure their effectiveness following 2017 when a decision on building new conventional generation must be made.

This multi-year pilot will study the performance of distributed generation (such as solar generation), energy conservation programs, and energy storage. SCE is working with key stakeholders such as business and residential customers and vendors to increase the use of these preferred resources in the region.

Because of the Preferred Resources Pilot, the company is likely to pursue a higher penetration of renewable generation in the area. According to Sergio Islas, Sr. Project Manager, Preferred Resources Pilot, "SCE has to overcome challenges with this project which include development in an area that is highly urbanized with space and land constraints."[7] Developers need to gain site control and overcome landlord/tenant concerns. From our vendors, we have learned that the minimum bid quantity of 500 kW in our initial solicitation was a market barrier to smaller firms. The bidding requirements were modified to allow floor bids of 250 kW so bidders could aggregate resources and bid in. The solicitation was completed in late 2015 with contracts for 2.17 MW of renewable generation—which illustrates some of the challenges with siting renewable generation in a highly urban area. SCE continues to work with various vendors to solicit input on how best to increase the adoption of preferred resources. While incorporating lessons learned and insights from potential vendors, SCE launched a second PRP-specific solicitation in late 2015 for 100 MW of preferred resources.

SCE is working to obtain the right mix of these preferred resources, that can meet the increasing demand for electricity in this region, which will reduce or eliminate the need to build local new gas-fired plants. Results of the pilot will be used to inform the clean energy grid of the future throughout our service territory and possibly the utility industry.

---

[7]Personal Communication, April 10, 2015.

## 12.11 Lessons Learned

The key lesson learned over the years is the importance of developing a collaborative model to renewables development, while at the same time protecting the customer's interest through contracting provisions. In addition to developing a collaboration with our developers, we have found it is critical to also consider and involve local, state, and federal agency stakeholders in the process, bringing them in early and providing updates as frequently as needed so that permitting and regulatory approvals do not derail the timing or cost of the project. Non-governmental organizations (NGOs) representing community and environmental interests also have concerns that need to be resolved as part of the process.

At the same time, one cannot discount the need for the utility to protect itself and its customers—as the utility is ultimately accountable for the delivery of energy from projects in its renewables portfolio, and customers incur the costs of these contracts. Bill Walsh, who has held various procurement roles at SCE for the past decade, notes that the company and market needs are constantly evolving. "Our contracts have undergone significant modification over the years, leveraging what already existed in the market, learning from past experiences, and changing contract provisions to support new and updated regulatory provisions."[8]

In addition, SCE has learned valuable lessons over the years in right-sizing projects, pre-screening developer bids, early consideration of transmission availability, and environmental impacts (*see* Challenges to Meeting Renewables Targets section above).

## 12.12 Conclusion

SCE sees competitive pressures, new technologies, and aggressive environmental policies where the energy sector is expected to contribute significantly to the reduction of greenhouse gases as prominent drivers of industry change in the coming years. The company will continue to adapt its business strategy to navigate through these changes, leveraging its experience and skill in pioneering new technologies to re-envision a flexible, resilient grid that enables customer choice and clean, sustainable energy.

---

[8]Personal Communication, May 26, 2015.

# Chapter 13
# Tennessee Valley Authority: Balancing Aquatic Biodiversity, River Management, and Power Generation

Tiffany Foster, Monte Lee Matthews, David Matthews, Hill Henry and Shannon O'Quinn

**Abstract** While operating an integrated system of dams, reservoirs, and power plants along the Tennessee River, TVA sustains a diverse ecosystem. The system is designed to provide for navigation, flood damage reduction, and the production of electricity. The TVA system encompasses more than 11,000 miles of shoreline and 650,000 surface acres of water within the Tennessee River watershed. The watershed covers approximately 41,000 square miles, with about 42,000 miles of streams and rivers. There are more aquatic species diversity (such as fish, insects, mussels, snails, and other forms of life) in the Tennessee River system than there are anywhere else in North America. TVA also operates the dams for other purposes, such as protecting water quality and aquatic habitat, providing municipal and industrial water supplies, regulating flows to minimize the effects of effluents, including thermal discharges, and controlling flows and levels for various recreational uses. TVA strives to minimize the impacts to aquatic life in its public power system and has embraced sustainable practices since its beginning. Through the years, sustainability at TVA has focused on the environment, the valley economy, and societal benefits.

---

T. Foster · S. O'Quinn
Tennessee Valley Authority's Natural Resources, Knoxville, TN, USA

M.L. Matthews (✉) · H. Henry
Tennessee Valley Authority's Environment & Energy Policy, Knoxville, TN, USA
e-mail: mlmatthews@tva.gov

D. Matthews
Tennessee Valley Authority's Environmental Permitting & Compliance, Knoxville, TN, USA

© Springer International Publishing Switzerland 2016
J. Fox (ed.), *Sustainable Electricity*, DOI 10.1007/978-3-319-28953-3_13

## 13.1 Introduction

Tennessee Valley Authority (TVA) is the nation's largest public power provider serving over 9 million people throughout portions of seven states in the southeastern USA. Throughout its history, TVA has made decisions through a lens of sustainable performance to improve the quality of life in the Tennessee Valley. Established by the TVA Act of 1933, TVA's mission of service to the region is to provide affordable and reliable electric power, responsible stewardship of natural resources, and sustainable economic development. These aspects of TVA's mission are managed in concert, benefitting the residents of the Tennessee Valley region.

TVA's public power system includes a diverse mixture of generation sources which provide power to a service area covering 80,000 square miles (Fig. 13.1) through a network of 16,000 miles of transmission lines. TVA's power system is integrated with TVA's River system of 49 reservoirs and dams, including 11,000 miles of shoreline, 650,000 surface acres of water, and 293,000 acres of adjacent land. This diverse and robust system allows TVA to provide strong natural resource stewardship and reliable power at competitive rates, both of which foster economic development in the region.

TVA's stewardship of natural resources contributes significantly to the Tennessee River region. Management of the river system and associated lands provides flood control, water supply, navigation, hydropower production, and land

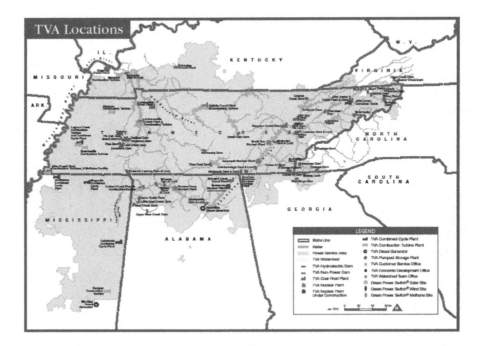

**Fig. 13.1** Map of the TVA power service area and Tennessee River watershed

and wildlife management and recreation. The integrated system fully contributes to the quality of life in the region by providing a diverse mix of recreation that fuels local economies. Brenda Brickhouse, TVA chief sustainability officer and VP Environment and Energy Policy notes, "TVA provides cleaner, reliable and affordable energy to support sustainable economic growth in the Tennessee Valley, and to engage in proactive environmental stewardship in a balanced and ecologically sound manner. TVA's pursuits in these areas serve the public to make life better for the people of the Tennessee Valley region."

To provide sustainable energy, TVA must consider the biodiversity of the region. The Tennessee River watershed covers approximately 41,000 square miles, with about 42,000 miles of streams and rivers. Considered one of the most biologically diverse regions in North America, the Tennessee River watershed is home to 230 native fish and 109 species of mussels, many of which only occur in this watershed. For context, the Upper Mississippi River watershed has 130 fish species, and approximately 300 species of mussels occur in North America. Fulfilling TVA's mission in a region with such high levels of biodiversity creates unique challenges and opportunities.

One unique challenge is due to TVA's role as a federal corporation. From 1933 through the 1950s, TVA received federal appropriations to fund operations. In 1959, Congress passed legislation making the TVA power system self-financing. However, TVA continued to receive appropriations for stewardship programs to supplement power revenue funds and non-power revenue sources such as agricultural leases, timber sales, and recreation management and user fees. With the passing of the Energy and Water Development Act in 1998, Congress ended TVA's federal appropriations and required TVA to fund "essential stewardship activities" including dam safety and maintenance, navigation, flood control, and natural resource programs. TVA is also expected to bear the costs of additional federal rules and policies; most federal agencies fund these costs through continued appropriations.

Incorporating funding of TVA's formerly appropriated stewardship responsibilities into TVA's self-financed integrated business model proved challenging. Essential stewardship activities competed with projects supporting critical power systems. TVA prioritized these activities focusing on human safety, minimizing flood damage, maintaining system integrity, and compliance with federal laws, regulations, and policies. Non-essential stewardship activities were reduced in scale or eliminated. As many provided positive benefits to the region, TVA engaged in collaborative efforts involving federal and state agencies, municipalities, and non-government organizations to collectively address beneficial programs valued by these stakeholders. Strong partnerships and collaboration are indeed a key component to TVA's stewardship success.

## 13.2 TVA's Approach to Stewardship

Sustainability is a broad concept that applies to TVA's mission and its role as a responsible natural resource steward. Brickhouse notes "Since its inception, the people of TVA have maintained a proud history of sustainable leadership to fulfill our mission of service." TVA's Environmental Policy recognizes how sustainability relates to TVA's mission:

> "The TVA mission is supported by its values, all of which reflect sustainability's social aspect: safety, diversity, integrity and respect, honest communication, accountability, teamwork, flexibility, and continuous improvement."

> "The TVA Environmental Policy and a commitment to provide cleaner energy correlates with the environmental aspect of sustainability. TVA's efforts to manage natural resources responsibly, reduce emissions, explore the use of renewable energy, all while providing affordable and reliable power, are central to this commitment."

> "TVA's economic development commitment mirrors the economic aspect of sustainability through goals of increasing capital investment and attracting and retaining good jobs for the people and businesses served by TVA."

TVA's management of the Tennessee River system incorporates this broad spectrum of sustainability through system-wide and site-specific programs. Some key initiatives and programs which have helped ensure TVA produce reliable power while safe-guarding biodiversity are outlined below.

## 13.3 A System-Wide Program to Support Aquatic Resources and Biodiversity

### 13.3.1 Reservoir Releases Improvement Program

As the nation's largest public power provider and steward of the nation's fifth largest river system, TVA operates a system of dams and reservoirs in an integrated manner to balance natural resource stewardship and power production with the other demands on the river system.

Prior to the Clean Water Act (CWA) of 1972, water pollution and sedimentation were major problems throughout the region. In passing the CWA, aquatic habitat conditions in the Tennessee Valley slowly improved. Additional improvements occurred when TVA's Reservoir Releases Improvement (RRI) program established thresholds for dissolved oxygen (DO) concentrations and minimum flows in TVA tailwaters. Because conditions differed at each dam, a variety of strategies were incorporated singly or collectively to improve DO concentrations and maintain minimum flows (Table 13.1).

Oxygen injection systems (Fig. 13.2) consisting of an oxygen tank and evaporators force gaseous oxygen through diffuser hoses suspended above the reservoir floor upstream of the dam. The oxygenated water is passed through the dam, improving DO levels in the tailwater. This system requires the routine delivery of liquid oxygen to remote distribution points.

Surface-water pumps (Fig. 13.3) push warm, oxygen-rich surface water downward, where it mixes with low oxygen on the bottom of the reservoir. The oxygenated water is drawn in by the turbines and passed into the tailwater, improving

**Table 13.1** Aeration systems used at TVA dams

| Aeration systems | Dam |
|---|---|
| Oxygen injection | Blue Ridge, Cherokee, Douglas, Fort Loudoun, Hiwassee, Norris, Nottely, Tims Ford, Watts Bar |
| Surface-water pumps | Cherokee, Douglas |
| Aeration weirs | Chatuge, South Holston, Norris |
| Air compressors and blowers | Nottely, Normandy, Tims Ford |
| Turbine venting | Apalachia, Boone, Cherokee, Douglas, Fontana, Hiwassee, Norris, South Holston, Watauga |

**Fig. 13.2** Oxygen injection systems are used to improve dissolved oxygen concentrations upstream and downstream of reservoir dams

water quality and aquatic habitat. The pumps are positioned above a dam's intakes and mounted on floats attached to a rail system, allowing the pumps to move up and down as the water level changes.

Aerating weirs are small dams designed to mimic a natural waterfall, adding oxygen to the water as it plunges over the top of the weir walls (Fig. 13.4). Aerating weirs are located a short distance downstream from dams. TVA has designed, built, and tested two types: a long W-shaped structure called a labyrinth weir that creates a waterfall, and a more compact structure called an infuser weir that uses a slotted decking to create a series of waterfalls. Weirs also serve to maintain wetted surface areas upstream of the weir and, as it drains, provide minimum flows downstream.

**Fig. 13.3** Surface-water pumps help oxygenate the water in deeper reservoirs, resulting in improved dissolved oxygen concentrations downstream

**Fig. 13.4** An infuser weir consists of decking made of metal grates. The water is oxygenated as it falls through small openings in the metal decking

Aerating turbine technology uses low-pressure areas to draw air into the water as power is being generated. At some dams, TVA has modified the existing turbines to draw air into the water. At other dams, TVA has installed new turbines specifically designed for this purpose. Hub baffles on the underside of a turbine cause air to be drawn into the water when the turbine is running (Fig. 13.5).

**Fig. 13.5** Hub baffles, cup-shaped structures surrounding the base of the turbine cone, inject oxygen into the water column while generating

### 13.3.2 Managing Water Flow

Changes in timing, duration, and magnitude of flow can influence DO concentrations and have a dramatic impact on biodiversity. TVA uses three different technologies, periodic turbine pulsing, weirs, and hydroelectric units to maintain water flow downstream of tributary dams. These techniques maintain minimum flows for a more constant wetted habitat downstream of the dams.

### 13.3.3 Response from Aquatic Communities

Aquatic communities are an important measure of RRI performance. Results from long-term assessments indicate that biological communities in most tailwaters are responding positively to the program. DO concentrations exceed target levels in 300 miles of tailwaters below TVA dams; minimum flows have improved in 180 miles of tailwaters. The number and diversity of fish and insects have increased in these improved zones.

Fish and aquatic macroinvertebrate communities in the French Broad River below Douglas Dam began such a dramatic comeback in the first five years of RRI activities that fisheries biologists began to believe that lake sturgeon, a native fish long since depleted from the region, could once again inhabit the area. A total of 13 upper Tennessee River tributary dams and two mainstream dams, all within the Upper Tennessee Management Unit, now employ some combination of aeration and operational changes to improve downstream releases. To date, nearly $70 million has been invested in TVA's RRI program.

### 13.3.4 Adjusting the Integrated System

TVA conducted a comprehensive Reservoir Operations Study (ROS) to determine the best strategy for operating the Tennessee River System while providing the best overall public value. Along with the US Army Corps of Engineers, and the US Fish and Wildlife Service (USFWS), we engaged an interagency team and a public stakeholder group to ensure that agencies and the public were actively and continuously involved throughout the study.

The three main areas of concern for aquatic resources were biodiversity, sport fisheries, and commercial fisheries. Environmental conditions (e.g., DO, water temperature, and flow) that potentially affected aquatic communities under the various policy alternatives were modeled and used as surrogates of aquatic populations and aquatic community responses. Predicted responses of aquatic resources were discussed at a programmatic level, and consideration was given to reduce negative impacts to aquatic life. The new operating policy was tested in both extremely dry and wet weather. In both cases, the system performed as intended, minimizing flood damage during periods of heavy rain, but also providing enough water for hydroelectric power and other downstream uses during dry weather (i.e., water supply, recreation). TVA was able to meet the main objectives associated with the new policy: water levels were held higher through Labor Day on most tributary reservoirs, and targeted recreational releases were met with very few exceptions.

### 13.3.5 Assessments of Streams, Tailwaters, and Reservoirs

A critical aspect of managing the Tennessee River system is accurate assessments of water resource conditions upon which we base sound management decisions. Aquatic communities support valuable commercial and recreational fisheries and are used by regulatory agencies to assess stream impairment. TVA performs multiple assessments within the Tennessee River system, including stream biomonitoring, tailwater ecological conditions assessments, major tributary biological and chemical assessments, and reservoir ecological health assessments.

TVA biologists adopted an innovative monitoring technique called the Index of Biotic Integrity to sample streams throughout the Tennessee Valley. Assessments support water resource programs that protect and restore water quality, aquatic ecology, and biodiversity in the Tennessee Valley and provide documentation for ambient conditions. Assessments are also used to prioritize hydrologic units for stream restoration projects, monitor stream restoration project success, and measure TVA's natural resource stewardship performance.

Stream biomonitoring is used to monitor ecological conditions in TVA tailwaters and assess the effectiveness of the Reservoir Releases Improvement Program and operational changes. TVA also monitors 18 major tributaries to the Tennessee River to determine the quality of water flowing out of each major watershed into the reservoir system. River assessments are based on quarterly water chemistry monitoring and biomonitoring. In addition to the stream monitoring, TVA monitors conditions in the reservoirs on the Tennessee River system. This information is used by TVA, stakeholders, federal, state, and local agencies and organizations to make reservoir management decisions.

## 13.4 Site-Specific Efforts to Support Aquatic Ecology and Biodiversity

### 13.4.1 TVA and the Elk River

In addition to system-wide efforts to balance dam operations with natural resources management, TVA has experienced success in several local efforts to protect and enhance aquatic ecology and biodiversity through system operations and species reintroduction. To comply with Endangered Species Act (ESA) requirements under Sect. 7 consultation, TVA modified operations at Tims Ford Dam in Franklin County, Tennessee to benefit endangered species in the 133-mile tailwater. At this facility, water releases are made through one hydroelectric unit, three spillway gates, and a low-level sluiceway. Forced air and oxygen injection are used to aerate releases from the dam to a target DO level of 6 mg/l. These measures help to improve downstream aquatic habitat.

Variability of flows and shifts in water temperatures downstream of Tims Ford Dam was impacting protected, warm water species, especially during summer months. Species of concern include the federally listed boulder darter (Fig. 13.6), and

**Fig. 13.6** Boulder darter (*Etheostoma wapiti*) known from the Elk River in Central Tennessee and North Alabama

**Table 13.2** Listed species known from the Elk River

| |
|---|
| • Boulder darter (*Etheostoma wapiti*) |
| • Birdwing pearlymussel (*Lemiox rimosus*) |
| • Cracking pearlymussel (*Hemistena lata*) |
| • Cumberland monkeyface (*Quadrula intermedia*) |
| • Fine-rayed pigtoe (*Fusconaia cuneolus*) |
| • Shiny pigtoe (*Fusconaia cor*) |
| • Slabside pearlymussel (*Pleuronaia dolabelloides*) |
| • Snuffbox (*Epioblasma triquetra*) |

numerous freshwater mussels, including birdwing pearlymussel, cracking pearlymussel, Cumberland monkeyface, fine-rayed pigtoe, rabbitsfoot, shiny pigtoe pearlymussel, slabside pearlymussel, and snuffbox (Table 13.2).

Adding to the complexity of operations in the system, a cold-water fishery was established immediately downstream of the dam, a popular resource with regional anglers. The juxtaposition of the needs of cold and warm-water species in the tailwater created an operational challenge for TVA. An adaptive solution incorporated the opposing needs of these resources and stakeholders, while ensuring TVA's regulatory compliance.

TVA uses a combination of sluicing and spilling incorporated through summer and fall to meet temperature and flow targets more closely resembling the natural free-flowing stream flows used by native mussel and fish species. This solution also maintained the cold-water fishery in the reach immediately downstream of the dam. Periods of hydrogeneration are resumed in late fall following reservoir "turnover" when water temperatures equalize upstream in Tims Ford Reservoir. These changes result in improved tailwater conditions while ensuring that summer reservoir levels of Tims Ford Reservoir (another variable important to recreation interests and landowners) are minimally affected.

Biological responses have been positive in both cold- and warm-water reaches. The slight increase in average water temperature in the upstream reach of the river improved species diversity, which increased prey diversity for trout. Recreational fishing opportunities have also increased in the cold-water reach.

The warm-water reach increased in length in response to operational changes, creating a potential recolonization zone of approximately 30 miles for sensitive mussel species. While little upstream movement by rare species has been detected at this point, multi-agency monitoring teams documented reproduction in several mussel species (Fig. 13.7). Multiple age classes of federally listed snuffbox and cracking pearlymussel have been recently observed at several sites in the warm-water zone. This positive outcome indicates aquatic conditions have improved in the lower Elk River.

With these notable improvements, state conservation agencies began reintroducing other imperiled mussel species into the tailwater. These introductions are proving successful; endangered Alabama Lampmussel, *Lampsilis virescens*, is exhibiting significant growth rates (Fig. 13.8) and good survivability.

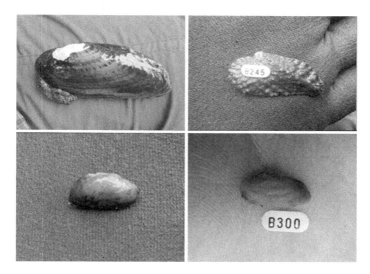

**Fig. 13.7** Recent recruitment of rare mussels in the lower Elk River after TVA modified flows in 2006. Clockwise from *upper left* adult cracking pearlymussel, juvenile Pistolgrip (*Tritogonia verrucosa*), juvenile cracking pearlymussel, and juvenile snuffbox

**Fig. 13.8** Positive growth rates of reintroduced juvenile alabama lampmussel (*Lampsilis virescens*) indicate aquatic conditions are improving in the lower Elk River. The *black arrow* shows new growth on mussels

TVA's adaptive management approach to managing flows at Tims Ford Dam is resulting in positive responses from aquatic resources while ensuring regulatory compliance with the ESA. However, while the adaptive approach strikes a balance between TVA's environmental and operational goals, it also comes at a cost in foregone revenue from reduced power generation during summer months at the facility.

**Fig. 13.9** Lake sturgeon, *Acipenser fulvescens*

## 13.4.2 Lake Sturgeon in the Tennessee River

Changes in water quality and management practices have also improved aquatic habitat and allowed successful reintroduction of other aquatic species elsewhere in the Tennessee River Valley. The lake sturgeon, *Acipenser fulvescens* (Fig. 13.9), historically occupied large rivers within the Cumberland and Tennessee systems; threats to the species included over fishing, impoundments, and deteriorating water quality. The last known records of lake sturgeon were from the French Broad River in 1900, Hiwassee River in 1942, and the Clinch River in 1967.

Biologists initiated efforts to reintroduce lake sturgeon in the Clinch River drainage in 1992 with a partnership between TVA, USFWS, Wisconsin Department of Natural Resources, and Tennessee Wildlife Resources Agency (TWRA). During the following two years, monitoring efforts failed to turn up any proof of survival. However, in 1996, an angler caught a 27-in. six-pound lake sturgeon in Norris Reservoir. Since then, there have been several unverified reports from anglers in Norris Reservoir. The species was later confirmed by biologists from Mississippi State University, who also caught several individuals in Norris Reservoir. While this initial project was only conducted for one year and was not considered adequate for reintroducing a sustainable population, it demonstrated the potential for successfully restoring the species in the Tennessee River Watershed.

Under the direction of the former Southeastern Aquatic Research Institute (SARI), presently the Tennessee Aquarium Conservation Institute (TNACI), a new multi-agency working group and recovery plan were proposed with the "primary goal of restoring a self-sustaining population of lake sturgeon to its historic range in the Tennessee River." To date, over 127,000 lake sturgeon have been stocked in the upper Tennessee River. Partners supporting this new recovery plan include TVA, USFWS, Wisconsin Department of Natural Resources, World Wildlife Fund, TWRA, Tennessee Aquarium, Tennessee Technological University, Tennessee Clean Water Network, University of Tennessee, United States Geological Society and Conservation Fisheries Inc.

## 13.5 Partnership and Public Outreach Efforts

TVA's Natural Resource Plan (NRP) focuses on the execution of strategic objectives outlined in the Environmental Policy. One focal area of the NRP includes water resource protection and public outreach. Implementing outreach campaigns to enhance awareness and appreciation of the region's aquatic biodiversity help support TVA's operational and programmatic efforts. Helping communities recognize the value of protecting aquatic species and their habitat, and what they can do to support protection efforts, is key to implementing long-term protection efforts. TVA's public outreach strategy includes collaboration with other federal and state agencies, municipalities, educational facilities, and non-government organizations to carry out environmental education campaigns and volunteer programs.

Successful outreach efforts often involve hands-on activities or outdoor demonstrations. Examples include Kids in the Creek, water conservation games, stream exploration trips, and sturgeon releases. Through Kids in the Creek programs, TVA biologists and partners offer hands-on educational opportunities to the public, with an emphasis on school-aged children. These efforts introduce children and adults to aquatic species in local streams and inform them of the importance of conserving water resources.

Outreach efforts have also engaged stakeholders to improve water resources throughout the Tennessee River Watershed. In 2014, TVA began working with 30 partners to protect five of the most biodiverse and vulnerable areas in the valley, including Bear Creek, Clinch/Powell, Duck, Elk, and Paint Rock watersheds. A comprehensive approach enabled TVA and its partners to implement water quality and habitat improvement projects for rivers that are home to an exceptional variety of fish and mussels, including many at-risk species.

Recently, TVA expanded its partnership work to examine potential aquatic ecology impacts related to climate change. The Southeast Monitoring Network monitors sentinel aquatic species to better understand potential biological, ecological, and hydrological responses of aquatic ecosystems. Data collected by the Southeast Monitoring Network and will be used to track conditions, conduct research, and support resource management decisions in light of climate change.

## 13.6 Kingston Recovery Project

In 2008, TVA experienced a dike failure at the Kingston Fossil Plant, releasing over five million cubic yards of ash into the Emory River, Watts Bar Reservoir, and onto adjacent lands. It was a challenging time for TVA and those impacted by the event, especially the local community. As the event unfolded, TVA management, engineers, biologists, and representatives of EPA, USFWS, TWRA, and other stakeholder groups gathered to collectively determine the environmental impact of this significant event. While teams worked to stabilize the site and assess potential impacts to the community, natural resource experts developed comprehensive studies to identify potential pathways for ash components to move through the ecosystem. Studies were performed at multiple levels across the food chain to better gauge resource impacts. These studies involved benthic invertebrates, amphibians, reptiles, birds, mammals, and fishes.

By the end of May 2010, TVA had taken all the necessary steps to protect public health and the environment, restore flow to the river, and minimize further movement of the ash. Even after reaching this milestone, environmental evaluations continued to further assess potential impacts to the ecosystem.

Over the course of the project, TVA has made significant investments in the redevelopment of the community and the surrounding environment. TVA's Kingston Ash Recovery Project received the EPA "Excellence in Site Reuse" award in 2015. At the public award ceremony, EPA said TVA's commitment to return the area to as good as or better condition was fulfilled. EPA's onsite coordinator noted that five years into the recovery, the river system's ecological conditions had returned to pre-spill conditions.

Collaboration between natural resource experts, regulators, and other stakeholders was critical to the recovery. Through this collaboration, new ecological assessment methods were developed and implemented. The data gathered under the new methodology helped TVA and regulators determine the ecological health of the impacted area and identify appropriate protection measures. The knowledge gained about ecological responses during the assessments helps TVA be better prepared to protect aquatic biodiversity and aquatic habitats, and enhances our ability to respond to future challenges. The ecological assessment methods, studies, and lessons learned from the Kingston Ash Recovery Project will prove valuable for TVA and others in the power industry who manage aquatic biodiversity, river management, and power generation.

Stemming from the Kingston event, TVA helped to create the Tennessee Healthy Watershed Initiative (THWI) using Supplemental Environmental Project (SEP) funds. This partnership kicked off in Fall 2011 when TVA, Tennessee Department of Environment and Conservation (TDEC), The Nature Conservancy (TNC), and West Tennessee River Basin Authority signed a Memorandum of Understanding and developed the THWI Charter to engage additional partners through this endeavor. In 2012, THWI requested proposals from organizations and communities across Tennessee from which nine projects were selected for implementation based on criteria developed by the THWI's Technical Advisory Group consisting of experts and scientists from TVA, TDEC, and TNC. Over 70 partners across the state have completed or are implementing 11 projects (two kick-off and nine grant projects) matching a THWI investment of $1,080,200 with $675,000. Due to THWI's initial success, $750K Environmental Restoration and Enhancement Project (EREP) funds were transferred from TVA to TDEC for THWI to fund a second request for proposal, bringing additional partners and environmental benefits to Tennessee. Currently, THWI is working with the EPA to implement the Healthy Watershed Assessment, which identifies healthy watersheds by assessing watershed conditions and future vulnerabilities. TVA stream monitoring data are provided as a valuable data set for this effort.

## 13.7 Conclusion

TVA has been actively involved with water resources and integrated management of the Tennessee River system since 1933, when Congress charged the agency with managing and serving as the steward of the Tennessee River and its watershed. Many of TVA's programs to date have been associated with construction and improvement of new dams and reservoirs and their operation as well as the broader stewardship mission of TVA. Strategic programmatic efforts including the ROS, Reservoir Releases Improvement Program, and the NRP continue to guide TVA's stewardship efforts to protect aquatic biodiversity while managing the river system and meeting stakeholder needs. As stewards of the Tennessee River watershed, one of the most highly biologically diverse watersheds in North America, TVA recognizes the need to protect these resources for future generations. Operating in such a rich, biologically diverse watershed presents challenges. However, through careful operational decisions and programmatic efforts, TVA continues to push forward managing the river system for a sustainable and diverse ecosystem while providing reliable energy and strong economic support, fulfilling its unique mission to improve the quality of life in the Tennessee Valley. Learning to strike the appropriate balance of TVA's energy production needs, stewardship mission, and aquatic biodiversity protection with often competing demands requires innovation, collaboration, and commitment. While maintaining the balance may not be easy, it is the right thing to do for the people of the region, the environment, and TVA.

# Chapter 14
# The Fourth Energy Wave: The Sustainable Consumer Is Knocking

**Clark W. Gellings**

**Abstract** America sharply redefined its non-transportation energy supply systems during three historical periods. Each period has changed not only how we provide energy but how we use it, with those forces working interactively. During each of these periods, there was a growing consumer awareness leading to the development of a fourth energy period in which the Sustainable Consumer resides. The first energy period was created by the infant electric energy industry of 1880–1910. The second energy period brought technological progress, consolidation, extension of electric service to most locations, and utility regulation. This era of progress, stability, and growth proceeded from 1910 through about 1970. The third energy period—from about 1965 through 2015—brought growing concerns about air, water, and soil pollution, as well as the potential dangers of nuclear power. While demands for "green" electricity have lagged behind some of these other initiatives, the electric power industry is unlikely to escape a robust response. In the future, consumers and suppliers will interact differently than they do today and bring dramatic changes that include an increasing focus on advancing sustainability. Sustainability is not just good business, it is the only business. Tomorrow's needs and the success of energy and energy services providers in the marketplace will depend on their response to the Sustainable Consumer.

---

C.W. Gellings (✉)
Electric Power Research Institute, Palo Alto, CA, USA
e-mail: clark.gellings@gmail.com

## 14.1 Energy Supply

The US economy relies heavily upon a safe, reliable, economical and increasingly environmentally responsible energy supply. Americans enjoy the fruits of a system of energy production, delivery, and use which is without peer, and use it heavily. Americans use energy for comfort and convenience; for production and distribution of goods and services; and for creation, processing, and distribution of information. Energy costs in the USA are low and stable, and energy supplies are admirably reliable. The system works well, with one glaring exception—it is unsustainable for future generations! Our production of electricity is not clean and as effective as it could be, our use of electricity is inefficient, we are wasteful in our use of water, and we generate an unacceptable stream of waste materials and pollutants.

The stakeholders groups which are involved in the electric power sector include a broad group of individuals and institutions, which are as follows:

- Produce, deliver, or otherwise provide electricity and electric energy services to residential, commercial, and industrial consumers;
- Provide equipment, technology, products, services, and systems needed to convert electricity to meet the consumers' needs;
- Providers of design, construction, operation, maintenance, and finance of such products and systems;
- Power systems operators as well as Federal, State, and local regulators of the electricity marketplace and its inhabitants; and
- All those who consume electricity (the consumers), either purchased or produced themselves.

*All of these stakeholders care about sustainability to one extent or another now—but, increasingly, it will be the consumer—The Emerging "Sustainable Consumer" that will drive the industry into a redefined future.* Each of the other stakeholders will operate enterprises and deliver energy, products, and services in response to the demands and desires of the "Sustainable Consumer."[1]

In three historical periods during the last century, America sharply redefined its non-transportation energy supply systems. Each period has changed not only how we provide energy but how we use it, with those forces working interactively. During each of these periods, there was a growing consumer awareness leading to the development of a fourth energy period in which the Sustainable Consumer resides. In the future, consumers and suppliers will interact differently than they do today, which brings dramatic changes that include an increasing focus on advancing sustainability.

The electric energy business is scarcely a century old. Born in an era of proud individualism, it has gone through successive periods of development triggered first by changing political and regulatory frameworks and then by unexpected external developments.

---

[1] Gellings and Gudger [1].

## 14.2 The First Energy Period (1880–1910)

The first energy period was created by the infant electric energy industry of 1880–1910. Utilities came about almost as an afterthought from the mind of inventor Thomas Edison. His main goal was to create a functional electric light bulb—a task that took him 8000 trials of different filament materials before he succeeded in 1879. Three years later, he threw a ceremonial switch to inaugurate electric service from the Pearl Street Generating Station in New York City. The nascent electric utility industry was born and immediately began fostering technological competition—alternating current versus direct current, ever-larger generators, and various prime energy sources—and competition for market share. Sustainability was almost solely about economic prosperity and, at that time, only incidentally about health and safety—such as the death of electrical workers caught in a web of overhead lines or which source of electricity was most preferred for human executions.

## 14.3 The Second Energy Period (1910–1970)

The second energy period brought order out of chaos left by the first. Inevitably, the growth created pressure for the combination of small electric supply and delivery companies into ever-larger enterprises. During this period there were increasing mergers of electric utilities, prompting public concerns of market power abuses from these new, larger entities. The early signs of issues of economic sustainability were born here, largely as a result of the public's positive response to successful government regulation on railroads. Americans turned to regulation as a means to control the excesses of the electric utility industry.

The second energy period brought technological progress, consolidation, and extension of service to all but the most remote locations. Its utility regulation brought the concepts of "obligation to serve" and "return on capital" as the yardstick for setting rates—a powerful incentive for utilities to encourage consumption. This era of progress, stability, and growth proceeded from 1910 through about 1970, interrupted only by the Great Depression and fueled by the Second World War. Utilities went for years without raising rates, and utility stocks were considered prime investments. Nuclear energy offered the promise of "too cheap to meter." The nation's energy system appeared immune to serious disruption. While polluted rivers and skylines darkened with soot were increasingly apparent, only a select few began to argue toward sustainability—that there was what Amory Lovins once referred to as a "Soft Path,"[2] one which focuses the electric sector toward sustainable solutions.

---

[2]Lovins [2].

## 14.4 The Third Energy Period (1965–2015)

The third energy period revealed that appearances were deceiving. The formation of the Organization of Petroleum Exporting Companies (OPECs) and subsequent oil embargoes triggered the infamous "energy crisis" in the USA and throughout the industrialized world. There were massive blackouts in the Northeastern US Raging inflation and 20 % interest rates ensued. Growing concern about air, water, and soil pollution focused on electricity production as the root of environmental evil. Nuclear power suffered the double blows of Three Mile Island (March 28, 1979)[3] and Chernobyl (April 26, 1986)[4], and the concerns over the environmental impacts of radiation were the subject of rock concerts, the first of several which was held in New York's Madison Square Garden and played to sold out crowds (September, 1979).[5] Of course, the increased awakening of society to issues surrounding sustainability was not just about the electricity sector. Society had become sensitive to sustainability decades before in actions such as the formation of the National Park system (August 25, 1916) or the Lewis and Clark Expedition (May 1804). There were many who campaigned for more careful consideration of our natural resources during this period, among the most notable was Rachel Caron who authored a book entitled Silent Spring (1962).[6] The environmental movement gained momentum throughout the 1970s and 1980s, and demands for "responsible" products emerged. By the 1990s, consumers began realizing the role of their purchasing power in influencing sustainable products. The emergence of the US Department of Agriculture (USDA) National Organic Program (NOP) (October 2002)[7] and the Forest Stewardship Council's Product Labeling Program (2004)[8] was a response to these demands.

While demands for "green" electricity have lagged behind some of these other initiatives, the electric power industry is unlikely to escape a robust response. In the 1970s, the industry was in an unsustainable economic situation with "revolving door" rate cases, and deregulation in various states, and it had started to become the focus of consumer concerns about the environment. Consumers increasingly understood that it would take heroic measures to clean the nation's air, its waterways, and species habitats and to stem the loss of species themselves. Each of the three pillars of sustainability, economic, social, and environmental, had been violated.[9]

---

[3]Backgrounder on Three Mile Island, www.nrc.gov.

[4]Chernobyl Accident, 1986, Appendix I, Sequence of Results, World Nuclear Association 2014.

[5]First of several sponsored by Musicians United for Safe Energy, www.beyondnuclear.org.

[6]Silent Spring, Rachel Caron, Mariner Books, Sept. 27, 1962, ISBN 0-618-24906-0.

[7]Oversight of the National Organic Program, Audit Report, March 2010, U.S. Department of Agriculture.

[8]Forest Stewardship Council Product Labeling, Bonn, Germany.

[9]Material Sustainability Issues for the North American Electric Power Industry: Results of Research with Electric Power Companies and Stakeholders in the United States and Canada, EPRI Report 3002000920, April, 2013.

Policy makers reacted to consumer outrage with an even heavier regulatory hand. President Jimmy Carter deemed the effort to reduce inefficient energy utilization "the moral equivalent of war,"[10] and signed legislation that helped create two new industries—demand-side management and renewable energy. Air quality regulations mandated emission controls. Safety enhancements were mandated on the nation's nuclear fleet. Many states embraced the process of centralized planning for electricity, then called integrated resource planning (IRP), for allowing equal consideration of demand-side management alongside new power plants. The challenges of the third energy period have led us to what we are now witnessing as the formation of the fourth energy period.

## 14.5 The Fourth Energy Period (2012–Future)

The fourth energy period is witnessing the emergence of the "Sustainable Consumer." This emerging historical period will be highlighted by a re-envisioning of the electricity sector driven by consumers with a heightened concern for sustainability. As described in Chap. 1, we are seeing an increasing number of sustainability-related shareholder resolutions, stakeholder protests, and performance disclosure demands. While these indeed are driving company action, ultimately it is the emergence of the direct consumer, the long-term purchaser of the product, which will require a wider response and restructuring of the systems, tools, and technology to ensure a sustainable product. Several basic recognitions are necessary in order to enable sustainability in the electricity sector[11]:

1. Recognize that electricity is a valuable energy form and much more than commodity energy; it is the underpinning of the modern quality of life and the nation's indispensable engine of prosperity and growth.
2. It is increasingly essential to enhance the value of electricity in a sustainable way in every element of the electricity enterprise, including generation, transmission, and distribution.
3. As the use of distributed energy resources (DERs), such as rooftop solar, expands (in particular distributed generation and storage), those resources should be integrated with central station resources.
4. As energy efficiency increases and carbon-free power generation expands, continued electrification is key to sustainability; most every end-use of fossil fuel in transportation, heating, cooling, and other uses should be replaced by electricity.
5. The Sustainable Consumer will become an active participant in, and benefactor of, the electricity enterprise.

---

[10]U.S. President Jimmy Carter, April 18, 1977, PD-USGov-POTUS, U.S. Presidential Speeches.

[11]Electricity Sector Framework for the Future-Volume I: Achieving the twenty-first century Transformation, Electric Power Research Institute, August 6, 2003.

Sustainability will require that the industry, the new entrants, and its regulators move beyond their traditional commodity culture in terms of electricity's value proposition to consumers and society. The Sustainable Consumer becomes a leader who embraces a vision that conveys the fact the electricity, through innovative technology, is a sustainable enterprise which provides a service value to society that is greater than its commodity value.

In part, the foundation for this exuberance in assuring sustainability comes from the evolution of innovative technologies which will enable the industry to meet the core mandate of providing safe, reliable, and affordable electricity while achieving goals in the areas of economic, social, and environmental sustainability.

## 14.6 Knowing the Sustainable Consumer

The *Sustainable Consumer* is characterized by a sincere interest in leaving this earth in a state that will ensure the well-being, health, and prosperity of our children and grandchildren. There are increasing signs that the Sustainable Consumer already exists and that through their actions and the growing popularity of such actions leveraged by the likes of social media, a growing awareness of climate change and political opinion polls, that this is not a passing fad.

In 2010, Yale University published an article on whether sustainability matters, being green in the delivery of products, and that service was identified as no longer differentiating products, but "it's expected. It's not differentiating."[12] The Yale paper went on to highlight how the trend toward sustainability will continue with tomorrow's consumer: "… within a very short period of time, the consumers will change. The younger customers, Generation Y—they are absolutely going to be voting with their dollars. They're going to be critical consumers." These customers have learned about sustainability in school "… and they are a completely different consumer than previous generations." A survey conducted by branding and marketing agency BBMG supports this hypothesis. It included a national sample of 2700 adults.[13] Ethnographers discovered five core values driving the more socially minded consumers:

1. Health and safety—unmodified organic products;
2. Honesty—companies must be accurate about practices;
3. Convenience—balance price with features;

---

[12]Does Sustainability Matter to Consumers?—From laundry detergent to automobiles, more and more businesses are presenting their products—and themselves—as green. How effective is green marketing? Will it have a meaningful impact on the planet? Yale Insights, Yale School of Management, May 2010.

[13]Conscious Consumers are changing the Rules of Marketing. Are you ready?—Highlights from the BBMG Conscious Consumer Report, R. Bemporad and M. Baranowski, November 2007. www.bbmg.com.

4. Relationships—who made it and where did it come from; and
5. Doing good—concern about the world and making a difference.

In BBMG's survey, they suggested consumers will provide a strong reward for social responsibility and sustainability by paying a fair price and pledging their loyalty.

In her 2011 book, The New Rules of Green Marketing,[14] Jacquelyn Ottman offered 20 new rules which include: "Green is mainstream," "Green is cool," and "Sustainability represents an important customer need, and is now an integral part of product quality." She went on the describe how these values guide consumer purchasing—that historically consumers bought on price, performance, and convenience—but, today how products [and services] are made, packaged, and disposed of and even how workers are treated, matter.

A similar study conducted by Penn State University[15] identified a growing segment of consumers which had a strong preference for environmentally friendly products and services. These consumers offered that it was important to them that they do not use products which harm the environment, and they consider the environmental impact in purchases and modify purchase decisions based on their concerns.

As with any strategic response, understanding the underlying values and core needs of the involved parties is critical for identifying a resolution that will be acceptable. Likewise, knowing the needs of the Sustainable Consumer is the first step the electric power industry needs to take to develop a responsive strategy—a strategy that will, by necessity, be complex and multifaceted.

## 14.7 The Sustainable Consumer and Electricity

The design of the sustainable power system must start with the consumer's needs and provide absolute confidence, convenience, and choice in the electric energy services provided so as to delight the customer. Perfection, based on the consumer's perspective, must be the design principle. Traditional electric sector planners, engineers, and scientists will find this perfection concept hard to embrace. Delivering sustainable electric energy services seems nearly impossible and inherently expensive—an impractical notion to those constrained by the conventional wisdom of the power sector. From the Sustainable Consumers' perspective, the sustainable electricity system has four attributes:

1. The first is its sustainability. It is essentially "green," being efficient in generation, delivery, and utilization of electricity.
2. It is functional.

---

[14]Ottman [3].
[15]Fryer [4].

3. Embedded in functionality is the reliable delivery of safe and secure electricity at the quantity and quality desired.
4. It provides high-valued energy services such that consumers can manage costs within their means while having the ability to exercise choice and connectivity.

The Sustainable Consumer will become a full partner in the electricity marketplace. Consumer services and choice will flourish in response to their demands as will an expanding array of electricity-based services that further increase electricity's value.

- The Sustainable Consumer will actively engage with their suppliers through the Web and other means to gain advice, and form and offer consensus on various sustainability issues. Power utilities and energy service providers will need to support these engagements.
- Consumers will receive price signals and participate in programs via smart appliances which reduce their consumption during peak periods without degrading services. Providers need to deliver these services.
- Consumers will employ their own generation and storage devices including photovoltaics and possibly microscale gas-fired and combined heat and power (CHP) generation. This will include facilitating the integration of plug-in electric vehicles with the power system. This will enable their participation in various retail and wholesale markets as well. Providers will offer these opportunities.
- Consumers will utilize some form of an "energy/information portal" to enable more efficient use of assets within buildings and industrial facilities, including participation in demand response programs, whether offered by the service provider or an aggregator. There are opportunities for providers to play active roles in offering these features to energy services.
- Service differentiation will become available for various classes of consumers offered by providers. In these offerings, consumers can avail themselves of differentiated electricity service by quality and price. These may range from premium digital-grade power to agreements on interruptible services.
- The Sustainable Consumer will become engaged in end-use innovation through their participation in buying and using various electricity-related energy services. Retail service differentiation will be extended into homes and businesses.
- The Sustainable Consumer will be poised to adopt technologies and services which increase electrification. With increasing performance of electric appliances and devices and the continued decline in the carbon content of kilowatt-hours generated, the Sustainable Consumer will be receptive to offers for electrification by service providers.[16]

---

[16]Sustainable Energy for All: Opportunities for the Utilities Industry, United Nations Global Compact and Accenture, Accenture. 2012.

## 14.8 The Sustainable Consumer Will Define the Future Power System

In a sustainable electricity sector, the Sustainable Consumer benefits directly and indirectly in a number of ways[17]:

- **Environmental and conservation benefits.** Sustainable energy service providers are basically "green." They reduce greenhouse gases (GHGs) and other pollutants by reducing generation from inefficient energy sources, support renewable energy sources, and enable the replacement of gasoline-powered vehicles with plug-in electric vehicles. They also reduce waste in power plants, employ energy efficiency throughout their own facilities, and heavily engage in promoting energy efficiency among their customer base. The sustainable power system will accommodate all generation and storage options. It supports large, centralized power plants as well as DER. DER may include system aggregators with an array of generation systems or a farmer with a windmill and some solar panels. The sustainable power system supports *all* generation options. The same is true of storage, and as storage technologies mature, they will be an integral part of the overall sustainability solution set. Renewable energy, increased efficiencies, and plug-in electric (PEV) support will reduce environmental costs, including the carbon footprint of the electricity sector.
- **Reliability and power quality.** In a sustainable world, energy service providers will provide more reliable energy, particularly during challenging emergency conditions, while managing their costs more effectively through efficiency and information. There will be fewer and briefer outages, "cleaner" power, and self-healing power systems, through the use of digital information, automated control, and autonomous systems. The future power system provides reliable power that is relatively interruption-free. The power is "clean," and the disturbances are minimal. Our global competitiveness demands relatively fault-free operation of the digital devices that power the productivity of our twenty-first-century economy. Society benefits from more reliable power for governmental services, businesses, and consumers sensitive to power outages. The future power system optimizes assets and operates efficiently. It applies current technologies to ensure the best use of assets. Assets operate and integrate well with other assets to maximize operational efficiency and reduce costs. Routine maintenance and self-health regulating abilities allow assets to operate longer with less human interaction. The future power system uses smart grids to independently identify and react to system disturbances and to perform mitigation efforts to correct them. It incorporates an engineering design that enables problems to be isolated,

---

[17]Power Delivery System of the Future: A Preliminary Estimate of Costs and Benefits, EPRI. Palo Alto, CA: 2004. 1011001

analyzed, and restored with little or no human interaction. It performs continuous predictive analysis to detect existing and future problems and initiate corrective actions. It will react quickly to electricity losses and optimize restoration exercises.
- **Energy efficiency benefits**. A sustainable power system is more efficient, providing reduced total energy use, reduced peak demand, reduced energy losses, and the ability to induce end-user use reduction instead of installing new generation in power system operations.
- **Direct financial benefits**. Sustainability offers direct economic benefits. Operations costs are reduced or avoided. Customers have pricing choices and access to energy information. Entrepreneurs accelerate technology introduction into the generation, distribution, storage, and coordination of energy.
- **Energy services**. In a sustainable world, energy service companies will enable consumers to balance their energy consumption with the real-time supply of energy. Variable pricing will provide consumer incentives to install their own infrastructure that supports the smart grid. Smart grid information infrastructure will support additional services not available today. The smart grid motivates and includes customers, who are an integral part of the electric power system. The smart grid consumer is informed, modifying the way they use and purchase electricity. They have choices, incentives, and disincentives to modify their purchasing patterns and behavior. These choices help drive new technologies and markets.
- **Operate resiliently to attack and natural disaster**. The future power system uses a smart and strong grid with hardened generation assets which resists attacks on both the physical infrastructure (power plants, substations, poles, transformers, etc.) and the cyber-structure (markets, systems, software, communications). Sensors, cameras, automated switches, and intelligence are built into the infrastructure to observe, react, and alert when threats are recognized within the system. The system is resilient and incorporates self-healing technologies to resist and react to natural disasters. Constant monitoring and self-testing are conducted against the system to mitigate malware and hackers. Providers continuously monitor themselves to detect unsafe or insecure situations that could detract from its high reliability and safe operation. Higher cyber security is built into all systems and operations including physical plant monitoring, cyber security, and privacy protection of all users and customers.

A benefit to any one of tomorrow's stakeholders can in turn benefit the others. The same changes that benefit utilites by reducing costs often also lower prices or prevent price increases to customers. Lower costs and decreased infrastructure requirements ameliorate social justice concerns around energy to society. Reduced costs increase economic activity, which benefits society. Benefits in the world of the Sustainable Consumer can be indirect and hard to quantify, but cannot be overlooked.

The future power system will enable a market system that provides cost–benefit trade-offs to the Sustainable Consumer by creating opportunities to bid for competing services. As much as possible, regulators, aggregators, operators, and consumers can modify the rules of business to create opportunity against market conditions. A flexible, rugged market infrastructure exists to ensure continuous electric service and reliability, while also providing profit or cost reduction opportunities for market participants. Innovative products and services provide third-party vendors with opportunities to create market penetration opportunities and consumers with choices and clever tools for managing their electricity costs and usage.

## 14.9 Making the Change

It is the variety of energy and energy service providers which will deliver these benefits and incorporate an expanding array of new technologies laced with sensors, communications, and computational ability. All of this added complexity to the traditional business of electric utilities is to be embraced, while the industry itself is undergoing massive change. Historically, the major players in providing electricity were those owning and operating generation, transmission, and distribution utilities. In many cases, these entities coexist in the form of vertically integrated utilities. Some of these entities are evolving by shedding groups of assets by spinning off generation or transmission, sometimes engaging in mergers or acquisitions. Foreign utilities are buying up some of the US players as well. Generation providers are now consumers themselves or new entrants usually referred to as independent or competitive power producers. A new breed of transmission owner–operator has emerged which combines the assets of traditional players into independent transmission companies.

Electric energy storage is evolving and becoming part of the power system, resulting in a change in generation and transmission requirements. This technology is expanding in effectiveness and being deployed by consumers, new entrants, and utilities themselves. Storage can augment generation, particularly to balance non-dispatchable generation, and it can reduce the congestion on delivery systems. There will be an explosion in the use of sensors and connectivity brought forward by a range of providers. Some of these information technology applications will provide the Sustainable Consumer the enhanced energy service opportunities already described by offering them greater control of their energy use, generation, storage, and plug-in electric vehicles as well as yet-unknown appliances.

The Sustainable Consumer and its buying power are important to these energy service providers for 5 key reasons:

1. **Revenues**—Sustainability presents opportunities for new businesses from both new entrants and existing players by expanding the array of new products and services or by enabling entry into new markets.
2. **Cost reduction**—Sustainability presents opportunities to decrease costs within energy service companies in operations and procurement. Maintaining a work force that adheres to a strict safety culture and which is satisfied with their work and the work environment.

3. **Risk reduction**—Sustainability includes the management of regulatory, environmental, and operations risk. Sustainability prevents costly accidents and avoids suits from the public on the impact of operations on health. It assures the long-term availability of water for electric generation and other uses. In addition, by assuring transparency in decision processes which involve major system expansion, the risk of subsequent course correction is minimized.
4. **Brand management**—Sustainability strengthens corporate recognition, reputation, the corporate brand, and stakeholder engagement. This is aided by contributions by energy service providers to their communities through procurement decisions, philanthropy, and volunteerism.
5. **Consumer satisfaction**—A sustainable energy service company will have a higher customer satisfaction rating than those companies who are less sustainable.

## 14.10 Barriers to Responding to the Sustainable Consumer

There are numerous barriers to sustainability in the near term. However, as the Sustainable Consumer evolves and we mature into the fourth energy period, most of these barriers will be overcome. Electric energy and energy service companies are unique in society when it comes to these barriers. In the USA, most electricity providers are currently regulated—in whole or in part by State and Federal regulators for expenditures, price, safety, environmental protection, market structure, and securitization, among other elements of their activities. Rather than detailing the barriers that presently exist toward sustainability in this sector, the focus here is on the medium to long term in order to identify the key barrier which is most likely to remain in the decades ahead. The assumption is that the demonstrated preferences and behavior of the Sustainable Consumer will overcome the lack of any business drivers which currently exist, leaving the key barrier affecting this sector in the medium to long term that of social and behavioral[18] attributes. Social and behavioral attributes are the most fundamental of all barriers toward achieving sustainability. This barrier includes those both those embedded in energy companies and those carried by certain groups of consumers themselves. Barriers within energy companies may stem from a lack of leadership. Energy company executives must embody leadership and have the political will to support sustainability despite their sometimes marginal return on investment and occasional opposition from those who want to maintain the "status quo." The utility industry is typically known for this attitude of "we have always done it this way!" A clear vision of a sustainable future needs to be part of everyone's DNA. Sometimes, utility internal actions are the largest barrier. Often, these stem from silos and incentives wherein counter-productive rewards and incentives as well as the presence of a counter-sustainability culture which believes sustainability is more work, more trouble, and more complex and costs too much is a deal killer in

---

[18]Jennie Lynn Moore, http://www.newcity.ca/Pages/what_stops_sustainability.pdf.

establishing a productive sustainability culture. For example, improved environmental sustainability must be valued in internal capital and budget allocation decisions.

As described above, the social and behavioral attributes of consumers are actually driving the fourth wave. It is the ongoing growth in consumers' preference for sustainability that is the basis for the Sustainable Consumer. The Sustainable Consumer's actions regarding the purchase of electricity, its use, and the purchase and operation of electricity consuming devices and appliances impact the environment and overall sustainability.

In order to understand the Sustainable Consumer's view of the electricity purchase decision and its consumption, it is helpful to examine an overall model for consumer decision making. The general decision process used by consumers in purchasing products is often outlined in consumer behavior theory as having four major stages:

1. Problem recognition (unmet need),
2. Alternative search and evaluation,
3. Actions [purchase(s)], and
4. Post action behavior (satisfaction?).

The decision process regarding energy purchases starts with recognition by consumers that changes in intrapersonal or external factors have evolved that result in discomfort or dissatisfaction. These in turn cause a heightened attention or awareness and prompt the consumer to begin decisions which the consumer hopes will rectify the discomfort or dissatisfaction. While uncomfortable or otherwise dissatisfied, the consumer becomes more receptive to, and may actively search for, internal and external information in an attempt to understand the "problem." This is a key point in the consumer decision process where information on sustainability can be critical to heightening understanding. Education at every level in today's educational and corporate system needs to begin today (if not already underway) so as to satisfy the consumer's search and helps to instill sustainability in all their actions. This includes an understanding that sustainability is also about economics—a sustainable energy and energy service industry is a profitable industry. Problem recognition, the first phase in the decision process, occurs when the consumer believes there is a significant difference between the extent to which perceived or actual needs are fulfilled and the fulfillment level desired or expected. To the extent that sustainability is a tenant in that problem recognition process will dictate how well consumers' lack of information is rectified by embedding sustainability into their understanding.

## 14.11 How to Accelerate Consumer Movement Toward Sustainability

Recognizing the importance of providing critical information on sustainability at exactly the right time so as to enable the fourth wave, there are several ways in which energy companies can touch their customers. The key question is how best to influence consumers in the selection of the best method(s) to obtain the desired customer acceptance and response and to convert all customers into Sustainable Consumers.

Customer behavior is influenced by the demographic characteristics of the customer, income, knowledge, and awareness of the technologies and programs available, as well as attitudes and motivations and their environmental impact. Customers are also influenced by other external factors, such as economic conditions, prices, technology, regulation, rebates, and tax credits. Many energy suppliers and governments have relied on some form of customer education to promote sustainability. Brochures, bill inserts, information packets, clearinghouses, educational curricula, and direct mailings are widely used. Customer education is the most basic of the methods available that can be used to inform customers about the relative sustainability of products and services being offered and their benefits, and influence customer decisions to participate in certain programs or to exercise preferences for certain technologies. Often, the message embedded in such programs targets the concept that sustainability increases the value of service to the customer.

Consumer education can be accomplished by direct customer contact techniques or face-to-face communication between the customer and an energy supplier to encourage greater customer acceptance of sustainability technologies and programs. Energy suppliers have for some time employed marketing and customer service representatives to provide advice on appliance choice and operation, sizing of heating/cooling systems, lighting design, and even home economics. These same representatives can be educated to emphasize sustainability. Direct contact can be facilitated by the use of workshops or energy clinics that can include special sessions that may cover a variety of topics, such as home energy conservation, third-party financing, energy-efficient appliances, other demand-side technologies, and even sustainability itself.

In addition, the use of exhibits and displays in large public venues, including conferences, fairs, or large showrooms, can be very effective.

Another way to influence consumers is to enlist the help of the energy companies' trade allies. A trade ally is defined as any organization that can influence the transactions between the supplier and its customers. Key trade ally groups include home builders and contractors, local chapters of professional societies [e.g., The US American Society of Heating, Refrigeration and Air Conditioning Engineers (ASHRAE), The Illuminating Engineering Society of North America (IESNA), and The Institute of Electrical and Electronic Engineers (IEEE)], trade associations (e.g., local plumbing and electrical contractor associations), and associations representing wholesalers and retailers of appliances and energy-consuming devices.

In performing their diverse services, trade allies may significantly influence the consumer's technology choice and behavior. Trade allies can substantially assist in developing and implementing sustainability programs.

Energy suppliers have used a variety of advertising and promotional techniques to influence customers. Advertising uses various media to communicate a message to consumers in order to persuade them. Advertising media applicable to sustainable messages and programs include radio, television, magazines, newspapers, outdoor advertising, and point-of-purchase advertising. Advertising and promotion also have widespread applicability. A number of radio and TV spots have been developed to promote sustainability.

## 14.12 Conclusion

Sustainability is not just good business, it is the only business. Tomorrow's needs and the success of energy and energy services providers in the market place will depend on their response to the Sustainable Consumer. The Sustainable Consumer will drive tomorrow's requirements. Those providers best able to respond will survive and prosper.

## References

1. Gellings CW, Gudger K (1997) The emerging open market customer—market smart consumers, new suppliers, and new products will combine to shape the "fourth energy wave". Edison Electric Institute, Washington, DC
2. Lovins A (1976) Energy strategy: the road not taken? Foreign affairs
3. Ottman JA (2011) The new rules of green marketing: strategies, tools, and inspiration for sustainable branding. Greenleaf Publishing, Sheffield
4. Fryer V (2014) 'Green' scale helps predict how consumers buy environmentally friendly products. Penn State University, Smeal College of Business, 24 Apr

# Chapter 15
# The Next Decade of Sustainability Science

Jessica Fox

**Abstract** If being sustainable is good for everyone, it would not be a debate or argument. The idea that environmental and social outcomes are cobenefits of "smart" business is interesting, but not yet realized. *If* it was possible and *if* the best way to do business was analogous with universally positive social and environmental outcomes, companies would have already made this transition. Yet, progress has been stymied by the complexity of identifying specific actions that adequately balance economic, social, and environmental targets. This is especially true in the electric power industry where the variety of natural resource dependencies, business models, and demographics makes industry-level solutions non-universal and rarely simple. Usually, priorities need to be determined, trade-offs assessed, and compromises made between the community, customers, employees, investors, and the environment. These are the complexities that are frustratingly absent from the popularized corporate successes. While environmental and social benefits can be cobenefits of financial outcomes in the electric power industry, we need to know the "right" decisions when win–win–win outcomes are not clear—this is the gap in sustainability science that needs to be addressed.

Why does sustainability feel like a fight? When I read acclaimed books on corporate sustainability, it is easy to walk away with the feeling that we are in a boxing match. If being sustainable is good for everyone, as many papers and books argue, it would not be a debate or argument. Yet, on one side, we continue to see the Friedman-type arguments that the most responsible choice for a business is to maximize its own economic well-being (see Chap. 1). On the other side, Bob

---

J. Fox (✉)
Electric Power Research Institute, Palo Alto, CA, USA
e-mail: jfox@epri.com

Willard provides a decade of robust arguments that companies lagging in adopting sustainable practices "are missing business opportunities" and that "Friedman-esque anti-CSR[1] proclamations" are halting collective progress.[2]

If we are in a sustainability boxing match, we might imagine a CEO,[3] a parent,[4] and a salmon,[5] three simplified perspectives of the triple bottom line. The ring is full, diverse, and complicated. What round are we in now? Are we at the beginning of this fight or near the final bell? What happens when one winner prevails or, miserably, one falls? Is there a way to reconstruct this match so that more than one winner is crowned?

Everyone loves win-win outcomes; that perfect solution where the core needs of both sides are met, compromises are accepted, and resolution achieved. Win–win–win outcomes, as in the sustainability triple bottom line, are even more appealing: winning for the environment, the community, and the business. But, from the perspective of a conflict mediator, as the number of "winning" parties increases, the probability of everyone leaving the negotiation table happy decreases. In the complex area of sustainability, not only do the number of conflicts increase, but the likely concessions are more intense, making compromises difficult to both identify and then settle upon.

If I were the mediator at the proverbial sustainability negotiation table, I would clarify that "consensus" means "I can live with that" not, "I got everything I wanted!" I would check where the CEO, salmon, and parent could flex without giving up core needs. However, when the salmon cannot give up the water that the CEO needs to run his power plant, and the CEO cannot find another affordable water source, and the parent is not able to pay more for his power bills, meet-in-the-middle solutions may be absent. When the degree of flexing is not livable, resolutions are not realized and the prospect of a win-win, let alone a win-win-win outcome, may be elusive, hence the complexity of the triple bottom line.

The idea that environmental and social outcomes are cobenefits of "smart" business is interesting, but not yet realized. *If* it was possible and *if* the best way to do business was analogous with universally positive social and environmental outcomes, companies would have already made this transition. Yet, progress has been stymied by the complexity of identifying specific actions that adequately balance economic, social, and environmental targets. This is especially true in the electric power industry where the variety of natural resource dependencies, business models, and demographic realities makes industry-level solutions non-universal and rarely simple.

The sustainability business case that is clear for leading companies such as Unilever, Patagonia, and Seventh Generation does not seem to address the

---

[1] CSR is Corporate Social Responsibility.
[2] Willard [1].
[3] The CEO works to protect his company, his employees, and thereby the associated community.
[4] The parent works to protects her children, our next generation.
[5] The salmon represents nature and broader ecological conditions.

complexity, diversity, and regulatory conditions of the electric power industry—an industry where sustainable choices come with difficult compromises. What happens when running a power company with no environmental impacts means increased bills for families? Or when there are fundamental trade-offs at play, opposing advocacy groups demanding different decisions, or regulations that prevent CEOs from choosing the "best" sustainability decision? Even a motivated CEO committed to triple bottom-line outcomes is left wondering what decision should or even *can* be made. These choices are not all-or-nothing. Usually priorities need to be determined, trade-offs assessed, and compromises made between the community, customers, employees, investors, and the environment. These are the complexities that are frustratingly absent from the popularized corporate successes.

In October 2015, Gov. Jerry Brown signed a law requiring California electric power companies to provide 50 % of their electricity from renewable sources by 2030 (currently at 20 %) (SB 350). Among many other projects, an offshore wind farm is proposed in Morro Bay, CA to provide renewable energy for 300,000 homes. While there are no risks of oil spills, no emissions or air quality impacts, and limited impacts to human health, there are still environmental and social concerns regarding birds and bats colliding with turbines, disruptions to fishing areas, and unpalatable shadows cast on this beautiful section of the California coast.[6]

For electric power companies, their non-generation activities are dwarfed by their power generation and delivery footprints. Reducing office waste, employee travel, office energy consumption, and similar highly cited approaches that are the primary efforts of many other industries are the "easy" activities for electric power companies. So, let us talk about the hard choices—water use versus affordable power, wind turbines versus bird strikes, off-grid solar versus continuous power, and distributed energy versus efficient grid backup.

Few resources are available to assist even the most thoughtful CEOs who want to make good sustainability choices but need more clarity of the specific trade-offs, when they will hit the financial books, and when the longer term benefits will be realized. Hence, business leaders in complex industries have been lacking concrete tools, resources, and equations to inform their difficult path towards decisions that are not black and white. It seems that "sustainability science" could inform these decisions.

To date, sustainability science largely focuses on sustainable development—research to support the technological, financial, and social advancement of countries and societies. Much of the research under this science is long running, theoretical, focuses on developing countries, and difficult to apply in a North American boardroom. In the next decade, sustainability science needs to include *applied* tools that businesses can start using now to achieve goals and inform choices. "Trade-offs" will likely be at the center of a sustainable science future.

---

[6]A Sea of Change for Our Energy Future? Sunday Times, Bay Area News Group. Nov 8, 2015.

An example of science informing trade-offs might be embedded in the triple bottom-line concept and might point to why the concept cannot be more broadly realized. It seems that the concept does not adequately recognize the three timelines of when the bottom line is reconciled. Financially, companies will roll up analysis annually, but more likely quarterly as they continually track economic progress, profits and loss, spending, and revenues. In contrast, ecologists think in terms of decades, if not centuries, for the environmental bottom line. For ecologists, reconciliation occurs every 10 years, give or take 5–100 years depending on the natural resource. The social bottom line is arguably somewhere in between but certainly will occur several times within the lifetime of a single person (maybe every 5–10 years). With reconciliation happening at different intervals from once a quarter to every 100 years, this differential reveals a gap between the triple bottom-line *concept* and the reality of its boardroom *application*. While the concept has motivated very important action, the sophistication of the idea and the resulting corporate examples need to advance. It is this advancement, and the associated science, that we need to complete in the next decade.

As we use models to predict environmental impacts of particular human activities, it may be possible to use models to match up the various timelines of triple bottom-line reconciliation into present day. A combination of models across disciplines would be needed with the ability to be ecologically predictive, financially accurate, and socially healthy. Likely, a complex set of metrics or indicators would need to be defined to feed into the combined model. This would be a mammoth exercise, but fortunately a strong foundation has been laid by existing ecological, business, and social science.[7]

In 2014, EPRI's sustainability research began an extensive effort to consider metrics that the electric power industry can use to measure performance, establish goals, and compare with peers.[8] The research encountered great complexity while working to identify common metrics across the industry. Even after agreeing on reasonable metrics, proceeding to use the metrics raised deeper questions, including the boundaries for reporting. Are electric power companies responsible for the species impacts of purchased power—the kilowatt-hours purchased from others and delivered to customers? If we look at water and species, there is currently no universal best practice for electric companies to assess the impacts of purchased power. What is the right normalizing denominator for such a diverse industry when comparing greenhouse gas footprints? Options include: number of customers (which can be measured either by the number of electricity meters, people, or larger/small users), number of employees, annual corporate revenue, or simply a raw total $CO_2e$ emissions that does not account for company size, customer base, or geographic spread. If sustainability assessments are left for investors or the public to assess, they may inappropriately match denominators and numerators, like $CO_2e$ emissions with miles of transmission line. Further, many metrics look at past

---

[7]Valentinov [2], Cavender-Bares et al. [3].
[8]www.epri.com/sustainability.

# 15 The Next Decade of Sustainability Science

**Fig. 15.1** The journey that leads to the science of sustainability (Graphical drawing provided by http://www.nityawakhlu.com/)

performance rather than predicting future performance or accounting for the long-term viability of ecosystems and society.[9] These areas may need to be revisited as we proceed to identify the next generation of sustainability science.

Figure 15.1 illustrates the next decade of sustainability science:

(1) Many definitions of sustainability;
(2) Unique challenges of the electric power industry;
(3) Current research to define metrics and models in sustainability;
(4) Balancing choices that are not black and white; and
(5) Finally looking toward what we will advance in the next decade.

While the concept of "sustainable electricity" is interesting, it continues to be a difficult concept to define. Is the definition "appropriately balancing trade-offs to meet the core needs of the environment, society, and business"? This book provides important examples and real-life challenges with electricity that is environmentally neutral, socially acceptable, and investor attractive. While environmental and social benefits can be cobenefits of financial outcomes in the electric power industry, we need to better understand which decisions provide these cobenefits. What are the choices that lead to those agreeable outcomes? For the

---

[9]See updates on EPRI's sustainability research at www.epri.com/sustainability.

electric power industry, we need to know the "right" decisions when win-win-win outcomes are not clear—this is the gap in sustainability science that needs to be addressed.

We will never be able to agree on who gets to win—the fish, the CEO, or the parent. Everyone needs to win, and it will require innovative scenarios and the development of scientifically-based tools that companies can use. The next decade of sustainability science will rely deeply on collaboration between industry, companies, and creative minds to take the gift-wrapped package from the top shelf of theoretical brilliance, remove the satin bow, and hand a set of usable tools to CEOs who can actually use them.

# References

1. Willard B (2012) The new sustainability advantage. Seven business case benefits of a triple bottom line. New Society Publisher, Canada, p 4
2. Valentinov V (2014) The complexity–sustainability trade-off in Niklas Luhmann's social systems theory. Syst Res Behav Sci. doi:10.1002/sres.2146/full, http://onlinelibrary.wiley.com (Published on-line Oct 2012)
3. Cavender-Bares J, Polasky S, King E, Balvanera P (2015) A sustainability framework for assessing trade-offs in ecosystem services. Ecol Soc 20(1):17. doi:10.5751/ES-06917-200117

# Index

**A**
Adaptation, 136, 137, 139, 140, 143, 144, 146, 149, 150
Air quality, 75
Aquatic biodiversity, 237–239
Avoided emissions, 155

**B**
Business, 257–261
Business vitality, 75

**C**
Carbon emissions, 77, 78, 80
Clean energy, 75
Climate change, 134, 136, 138, 140, 148–150, 152, 158, 166, 212, 237
Collaboration, 13, 17, 20, 29
Combined heat and power, 55, 56, 62, 248
Communication, 8, 23, 24, 26
Communities, 198
Culture, 7–9, 16–18, 23, 25, 27–29
Customer service, 42

**D**
Demand-side management (DSM), 31–33, 35–43, 45, 47, 48, 51
Distributed energy resources (DER), 221, 249
Distributed generation, 54–56, 58, 61, 72, 74
DTE energy, 91
Duke energy, 111

**E**
Electric power, 2–4, 6
Electricity use intensity (EUI), 202
Employee engagement, 17–19, 22–24, 27, 112, 113, 115, 117, 122, 123, 128, 129, 130, 131
Energy conservation, 33, 42, 46, 254
Energy efficiency, 36, 41, 46, 48, 50, 86, 115, 119, 129, 152, 154–157, 159, 160, 162, 164, 183, 185–187, 192, 194, 196, 197, 199–202, 204–206, 208–210
Energy storage, 54, 73
Entergy, 133
Environmental movement, 244
Environmental stewardship, 112, 136, 186, 227
Exelon Corporation, 151

**F**
Financial performance, 4
Flood protection, 145
Fourth energy wave, 241
Future trends, 27

**G**
GHG goal, 153, 162
Global reporting initiative (GRI), 9, 14
Greenhouse gas emissions, 164, 200, 202, 207, 212, 216
Greenhouse gases, 249

© Springer International Publishing Switzerland 2016
J. Fox (ed.), *Sustainable Electricity*, DOI 10.1007/978-3-319-28953-3

## H
Habitat protection, 105, 109
Hoosier energy, 169
Hydroelectric dams, 231, 233
Hydroelectric power, 212, 215

## I
Infrastructure resiliency, 166
Integrated resource planning, 33, 43, 44, 49

## L
Leadership in energy and environmental design (LEED), 186

## M
Marketing, 36, 40, 45, 46, 51
Materiality, 13–16
Metrics, 260, 261
Models, 258, 260

## N
Net metering, 54–56, 59, 73
New energy economy, 76, 89
Non-governmental organizations (NGOs), 11
Non-investor-owned utilities (non-IOUs), 171
Nuclear generation, 152, 155, 166

## O
Once-through cooling, 94

## P
Photovoltaics, 55, 57, 216, 217, 248
Power generation, 225

## R
Renewable energy, 176–178, 192, 194, 195, 212, 213, 215, 216, 220, 222, 223
Renewable generation, 78

Renewable portfolio standard (RPS), 155, 157
Resiliency, 136, 138, 143, 145, 149, 150
Risk management, 138
River management, 238
Rural cooperatives, 180

## S
Smart grid, 67
Solar power, 54, 58, 67, 69, 81, 86, 87, 214
Southern California, 211
Stakeholder engagement, 8, 10–12, 17–19, 24, 27, 29
Storm hardening, 138
Sustainability, 2, 3, 5, 8–10, 12–14, 17, 20, 22, 23, 26, 27, 32, 38, 44, 50, 111–113, 115, 117, 119, 120, 122–124, 126–131
Sustainability science, 259, 261, 262
Sustainable consumer, 242, 245, 247–251, 253–255

## T
Tradeoffs, 2, 259, 261
Triple bottom line, 2, 32, 41, 50, 51, 170, 173–175, 179, 189

## W
Water conservation, 95, 103, 109, 197, 199, 205, 210
Water consumption, 92
Water-energy nexus, 99
Water management, 92, 104, 109
Water quality, 92, 97, 229, 232, 236, 237
Water withdrawal, 92–95, 109
Watersheds, 237, 239
Wetlands restoration, 140, 141, 146, 149, 150
Wind energy, 214
Wind power, 87